Corporate Networks:
The Strategic Use of Telecommunications

For a complete listing of the Artech House Telecommunciations Library,
turn to the back of this book

Corporate Networks:
The Strategic Use of Telecommunications

Thomas Valovic

Artech House
Boston • London

Library of Congress Cataloging-in-Publication Data

Valovic, Thomas
   Corporate networks ; the strategic use of telecommunications / Thomas Valovic.

   p.   cm.
   Includes bibliographical references and index.
   ISBN 0-89006-484-9
    1. Business—Communications systems—Data processing—Case studies.
2. Telecommunication systems—Case studies. 3. Corporations—united States—Communication
systems—Data processing—Case studies. 4. Computer networks—United States—Case studies.
I. Title.

HF5548.2V294   1992   92-19581
658.4'038—dc20   CIP

© 1993 ARTECH HOUSE, INC.
685 Canton Street
Norwood, MA 02062

International Standard Book Number: 0-89006-484-9
Library of Congress Catalog Card Number: 92-19581

10 9 8 7 6 5 4

To my wife Elaine and daughter Jennifer

Contents

Foreword

Readers, fasten your seatbelts. This book is going to make you think. Whereas most books on telecommunications seem to be enumerations of technologies and products, Valovic shows you the raw material, knits the fabric, and then goes on to stitch together garments of increasing sophistication. Your perspective on the importance of telecommunications and computing in a corporate environment may well change. At the very least, it will have left you with a final challenge: how to apply new insights to your own enterprise.

The confluence of telecommunications and computing and their transformation into both personal- and enterprise-enabling technologies lie at the heart of a new revolution. The intimate involvement of portable computing capacity in everyday business practice and the integration of software into the myriad transactions that fuel the interenterprise economy form the warp and weft of a new business fabric. Telecommunications services of all kinds (voice, video and data) are fundamental elements of this new fabric. Computing capacity, protocols, and software designed to exchange standardized data structures are basic building blocks from which complex intra- and interenterprise applications are constructed.

Two strong forces for change may be singled out for special consideration in accounting for the transformation of business practices in the 1990s: personal computing and computer networking. In simple terms, personal computing empowers an individual; computer networking does the same at an enterprise level.

What are some of the features of this new information infrastructure? The list that follows is neither organized nor complete (read the book for a much more thorough effort!), but it is intended to offer glimpses of what the French would call business informatique for the new century.

1. *Fluid work group formation*. Combinations of voice and videoteleconferencing, together with shared workspace (for example, multiparty documents, display spaces), all of which can be interlinked "on-the-fly," allow the forming

and reforming of many workgroups as the need arises. Some technologies such as e-mail and voice mail remove the strictures of time-zone simultaneity, but these cannot fully replace the interactive environment of concurrent group meetings, nor should they be expected to.

2. *Enterprises as information systems.* Joseph F. Coates is quoted wonderfully in Chapter 3: "Your enterprise is an information system which incidentally sells goods and services. . . ." This observation is as important within as well as between enterprises. The real power of computing and telecommunication integration shows itself only when the information management functions of the enterprise have become a natural part of the enterprise software environment. Software that has built-in task models containing knowledge of the jobs people must perform can make far more powerful use of computer networking and information standards to assist each user to do his or her job best.

3. *Electronic data interchange.* The primitive "anonymous FTP" capability of the Internet seems a long way from the promise of electronic data interchange, but one can see the seeds beginning to germinate. An anonymous FTP database is a kind of advertisement or catalog in that it is available to anyone who is interested. In the academic world, these databases typically contain research results, papers and raw data. In the business world they might contain product descriptions, price lists, and order forms. Only in the last year or so has the Internet community begun to explore ways to organize and help index these thousands of information sources. It seems predictable that once it becomes natural to find these sources of information, their use as business tools will follow. E-mail, enhanced with cryptographically supported privacy and authenticity, will form a solid basis for interenterprise communication.

4. *Flattening the communications hierarchy.* It is commonly thought that e-mail has led to a kind of flattening of the command structure of the enterprise. I think this is less true than folklore would have it. What has happened, however, is a fattening of the communications channels among various parts of the organization. In the past, the top levels of management had only a thin ability to communicate with the levels farther down; the addition of e-mail has made it possible for much broader interactions among all levels of corporate structure. The fluid work group phenomenon also allows much more diversity in the population of work groups, drawing individuals from many parts of the organization at all layers. There is still a corporate hierarchy, but the barriers between levels has become permeable for purposes of interaction and work group formation.

5. *Staying connected on the road.* Portability of computing and communication power makes it far more possible to stay connected with the office environment while traveling or even at home (one may argue over the value of the latter!). The rapid gain in processing power and local storage capacity, cou-

pled with increasingly available digital telecommunications services, will soon render moot the distinction between being on the road and being at the office.

6. *Feedback from the point of sale.* Valovic makes the observation that information about individual transactions is increasingly available as raw material with which to measure enterprise-related trends and conditions. The obvious possibility of linking orders from suppliers to data on sales in a fully automatic fashion gives way to more interesting possibilities such as detection of new trends, improved opportunity for customizing products and services to consumer interests, and even the resale of processed information to enterprises in related businesses. I believe that customization will become increasingly possible as manufacturing enterprises integrate computer tools into the process. Perhaps economies of scale will be replaced by economies of information—with enough data, even unique devices may be manufactured in a cost-efficient fashion.

7. *Plastic money and its electronic counterparts.* It has long been out of fashion to predict cashless economies. However, there is a growing trend toward treating electronic transactions as if they were cash. Debit cards are much more common now. Credit cards are showing up in grocery stores (replacing checks, which, to be fair, are not cash either). If critical mass is ever reached (that is, the bulk of any personal transactions being carried out electronically), then one can expect software products for analyzing and organizing the resulting transaction information (for example, for tax purposes, for analyzing personal cash flow, and so on). The decade of the 1990s appears to be at least as pregnant with change as the decade of the 1890s seems to have been. Our world models were transformed at the start of the twentieth century in ways we could not have predicted. It is entirely possible that the new world that awaits us in the twenty-first century is equally unpredictable. We have some idea of the elements that will go into its socioeconomic fabric, but the combinations are too varied to enumerate and the potential for innovation too high for simple projections. What one can say with equanimity is that what may seem today the most irresponsible speculation may prove to underestimate the reality that is to come.

Vinton G. Cerf
Annandale, Virginia
April 1992

Preface

This book was born of a curious mix of both frustration and enthusiasm. The frustration was my own personal disatisfaction with how a radically transforming technology has been misinterpreted and misunderstood, especially in the general business press and popular media. Business publications, with their necessary obsessions about unit shipments, annual revenues, and products that come and go seem all too often to miss the real significance associated with this fascinating new technology. That significance is, of course, the impact it has on every one of us as we go about our daily professional practice. Unfortunately, all too often computer and communications technology tends to be viewed through the filter of the same shopworn categorical descriptors that are used to frame the traditional smokestack industries. Business reporters often write about it as if they were talking about farm equipment, microwave ovens, or soap flakes. However, computer and communications technology is emphatically not just another item in the chain of endless consumer goods: it represents a powerful new tool for transforming the workplace and society.

That's the frustration part; but there's another side here, and an important one. As editor of *Telecommunications* magazine, I also get to see a lot of positive developments and applications across the industry, while talking with others who are as convinced as I am that, in conjunction with the right "mindware," networking technology can become a positive force for change, not only in the corporate environment (which is going through its own set of changes), but also as a means of coping with what seems to be a staggering array of economic and sociopolitical problems. There are unique and innovative uses of this technology being implemented every day, and companies are using computer networks to shake up their business and reinvent the meaning of both competition and cooperation. There are forward-thinking managers who, understanding the nuances of the word software, view the networked computer as a malleable tool of theoretically infinite variety. From a McLuhanesque perspective, computers are extensions of our own minds, and, as such, they embody great potential for personal empowerment. But for this to happen, users need to be given the requisite tools and understanding to get them interested in exploring. And that, in my mind, is the real key. If desktop users of computer and communications tools are allowed to approach this new technology playfully and creatively, then they will indeed explore. (A touch of the Zen concept of "beginner's mind" can be helpful here!) Such exploration and the curiosity that propels it will then allow new realms of creativity to permeate even the most humdrum professional tasks. In the best sense, then, work and play merge. I remember a conversation with Howard Salwen, the chairman of Proteon, who described the computer as if it were a musical instrument that users could eventually "play," with swiftness and agility. Salwen's image has stayed in my mind ever since. It isn't such a far-fetched analogy.

This book was the culmination of many musings over an extended period of time concerning a technology that I've observed and reported on for close to ten years. The way it all came together was, however, different than what I anticipated. I remember reading a description of how Ithiel de Sola Pool wrote the classic text *Technologies of Freedom* by first jotting down bits and pieces about various topics and then weaving them together, orchestrating the whole into more than the sum of its parts. I also remember thinking at the time: what an odd way to write a book. Ironically, however, I seem to have done much the same. Editors continously encounter all sorts of stray pieces of information, and this, rather serendipitously, seems to have been a major factor in shaping the organizational dynamic of this book. In many cases, I didn't have to go out hunting for research material; the research material came to me, and I merely tried as best I could to grab the brass ring. Accordingly, *Corporate Networks* is essentially a mosaic of many sources of information that often presented themselves in the unrelenting whitewater of an information stream that is the editor's natural habitat. More importantly, I hope that readers of this book will find the results interesting and worthwhile.

In surveying the landscape of available texts on the strategic use of telecommunications, I noted that many books tended toward one of two extremes. The emphasis generally seemed to fall either on the subject of the technology itself, or else on the applications that they made possible. In writing this book, I wanted to not only avoid those extremes, but also to take pains to show relationships between them: how do technology and associated applications fit together and how are they mutually reinforcing? I also attempted to paint a broad picture so that the material presented would not disadvantage nontechnical managers unfamiliar with specific technologies. By the same token, for those with a more technical orientation, I've tried to provide useful information about the vector dynamics of macrolevel trends that will define the industry for some years to come. In trying to fulfill these goals, the objective was, as much as possible, to place these emerging trends in the context of some of the new business organization and management theories being developed by people like Harvard economist Robert Reich in his excellent book *The Work of Nations*, Alvin Toffler in his less scholarly but well-written and thought-provoking book *Powershift*, and Charles Savage in *Fifth-Generation Management*, who manages to make a convincing case for a kind of parallel evolution of modern management practice and the technical development of the computer through five successive generations. Using these theories as background, the book seeks to explore how computer and communications technology has a fundamental effect on information flow within the corporation, and how that information flow, in turn, shapes the very nature of the way companies structure themselves.

Finally, we no longer dwell in a linear cultural environment. Accordingly, I've tried not to create a linear text, but rather to make each chapter modular and somewhat discrete. Thus, readers are encouraged to sample those areas that they are most interested in. Nontechnical readers might wish to skip Chapter 2, which tends to be fairly advanced for those unfamiliar with the core technologies mentioned.

I'd like to thank a number of colleagues and associates who have helped with this effort, both directly and indirectly. First of all, thanks to William Bazzy and William M. Bazzy—chief executive officer and president of Horizon House, respectively—for fostering the book's development during an extremely busy business cycle; to Vinton Cerf, president of the Internet Society and reviewer for the developing manuscript, for his feedback, comments and refreshingly unambiguous opinions; to Artech editor Mark Walsh, who, had he not encouraged me to take on this subject matter, might have ended up with a book on personal communications networks. Mark also provided some helpful suggestions and recommendations during some critical turning points in the book's development; to Charlie White, former executive editor of *Telecommunications*, who for five years has been a mentor and guide through the complicated tangle of companies, technologies, markets, and applications that is the communications industry; to Tony Rutkowski, a founding member and vice president of the Internet Society and a former senior officer with the International Telecommunications Union in Geneva, for encouraging my explorations into the more complex dynamics of the industry from the beginnings of our short but enjoyable association; to Jim Budwey, for his helpful perspective in reviewing and commenting on Chapter 8; to Linda Garcia, project director with the Congressional Office of Technology Assessment, who provided valuable feedback and insight concerning some of my more far-reaching theories about major trends in the industry; to Peter Corr, who reviewed Chapters 2 and 8; to Pam Ahl of Artech, who was a source of cheerful and practical encouragement over the last year and a half and was always helpful in answering questions; to Rebecca Warren, whose easygoing style smoothed over some early difficulties in the early stages of the manuscript; to the other Artech staffers involved in the development of this book, both directly and indirectly; and to many other knowledgeable professionals I've met in the course of my years as an editor, whom I am constrained by space from mentioning.

And, finally, a special note of thanks to my wife Elaine—to whom this book is in part dedicated—for her patience, encouragement, and helpful advice throughout what sometimes seemed like an endless project.

Thomas S. Valovic
Needham, Massachusetts
April 1992

Chapter 1
Introduction

Only connect.

—*E. M. Forster*

Most revolutions occur quietly. A single individual toiling over a laptop computer hardly looks like the foundation for radical or revolutionary change. But look more closely: this person is connected via a modem to a worldwide network of knowledge bases, many of them the most important repositories of scientific and technical learning in the world. Making connections at the speed of light, this person can rove effortlessly through the libraries of major European universities, accessing information on a wide variety of subjects. What in the past would have taken years of travel and sheer scholarly determination now transpires with a few taps on the keyboard or clicks of the hand-controlled mouse. This is the unparalleled power of computer networking.

But what's wrong with this picture? Look again and you'll notice that such people are not necessarily employed by a major corporation. Nor are they affiliated with a library, university, or institute of higher research. Individuals who explore these on-line labyrinths do so under their own sponsorship. Along with other similar adventurers, they do this so-called "information-surfing" in large measure *because* they know no one else does. They also know that no federal grants, corporate research projects, or other institutional mechanisms are likely to support their efforts. They and their colleagues have recognized the new power of this medium and have decided to discover its potential with or without the help of sponsorship. In some quarters, these types of explorers are ungraciously dubbed "hackers," a term that is sometimes pejoratively used to describe those who gain unlawful access to others' computer systems. However, to characterize the majority of such individuals this way would be grossly unfair and inaccurate; by and large they are working for positive goals. In fact, what they are doing is providing the model for the way that people in corporations and many other types of institutions will work in the not-too-distant future.

The message is clear: as a society, we are not even close to utilizing the inherent potential of computer and communications technology. Furthermore, in one of the areas in which the United States most needs this to happen, the corporate environment, progress is made at a disappointing pace. Yes, computer networking in all its forms and expressions is indeed a quiet revolution; so quiet in fact, that many of our business and political leaders have yet to take full notice.

This is not to say, of course, that many corporations weren't heavily involved years ago in the purchase of those lumbering beasts called mainframes. These large systems formed the basis of corporate computing for many years and relegated the power of the computer to a kind of glorified bean-counting: administrative tasks, personnel statistics, budget crunching and so on. The computer resided in a special room, surrounded by glass, that outsiders often visited with all the hushed reverence that might be expected if it were a hospital operating room. These multibillion-dollar systems were fussed over and ministered to by a priestly class of computer experts who spoke computer languages—COBOL, FORTRAN, PASCAL, C—more readily than English. But let there be no mistake: the traditional IBM-based, big-systems "glass house" was *not* to be the real scene of this particular revolution—not by a long shot. In fact, what this revolution was all about was the shattering of the crystalline elitism of the glass house, the liberation of the computer from the confines of the big-systems experts, and its descent into the hands of ordinary workers who have now begun to use them to accomplish extraordinary things.

Why is the corporate world still ambivalently struggling to embrace this new revolution that empowers its workers and affords striking competitive advantages? The answer is a complex one. In deference to that complexity, I've attempted to piece together a kind of mosaic of observations and descriptions that also address the often-ignored cultural issues in the workplace. It's my hope that the reader will come away with a sense of the obstacles as well as epiphanies associated with the gradual introduction of computer and communications capability in the last thirty years. But the short answer is simply this: for mainstream corporate organizations—especially large companies—the paradox of a very unpleasant tradeoff was built into decisions about how to implement the new technology. The tradeoff was this: computer networking offered companies radical new possibilities for sustaining competitive advantage, but the price of these new capabilities was often nothing less than the fundamental restructuring and redesign of the corporate organization itself. Needless to say, some companies deemed this too high a price to pay and, with or without full management intent, elected to place the deployment of these new technologies on the slow track. In many cases, however, the net effect of such strategies was simply to defer this restructuring to a later date when forces even more radical, quick, and sure—namely economic corrections of one sort or another—would converge to create similar changes in organizational structure, only in a more painful fashion.

Despite the slow rate of acceptance of these technologies by some companies, there is no question that progress is being made and important changes are taking place. Many corporations have indeed moved boldly ahead to embrace new strategic approaches to the marketplace based on emerging technologies and, in doing so, have become experimenters and pioneers in their use. America's much-discussed competitive problems are by no means close to being solved, however, and it seems clear that such changes—implemented earlier and more proactively—might have helped U.S. industry avoid the economic doldrums that occurred during the difficult period of the early 1990s. Of course, more enlightened use and deployment of computer networking shouldn't be construed as a panacea; still, it has been significantly underestimated by both business and government leaders as a means of making U.S. companies more resilient and competitive in the new and emerging global marketplace. Evidence of this unfortunate fact of public-policy dialogue can been seen in the fact that discussion of such items as the National Research and Education Network (NREN)—championed by Senator Albert Gore—rarely occurred during the presidential campaign of 1992. The subject of America's global competitiveness was often mentioned, as was the the need for better education and new approaches to the problems associated with the state of U.S. manufacturing. Nonetheless, few of the candidates had the vision and foresight to bring computer and communications technology into the limelight of public debate, where its potential positive impact on these other critical matters could be fully explored [1].

AMERICAN BUSINESS: PLAYING CATCH-UP?

Good news must travel slowly when it comes to technical invention! There is a story that many have undoubtedly heard in one forum or another and that, apocryphal or otherwise, strikes a resonant chord. When Alexander Graham Bell presented the results of his invention—the telephone—to one of the great captains of American industry, Bell explained how, through the use of the device, an individual in one part of the country might eventually be able to converse with another located hundreds of miles away. The executive was said to have responded by asking, "Yes, but why would anyone want to do that?"

In these days of advanced telephony services, we tend to chuckle knowingly at this story, portraying, as it does, a mind-set that we perceive as unprogressive, atavistic, or socioculturally backward. Unfortunately, remnants of this kind of thinking are quite evident in the corporate boardrooms of many U.S. corporations. The only difference is that this time, the technological capability that's being questioned is not the telephone, but one that represents probably the most radical transforming principle ever to be introduced into a business or societal context: the potent combination of computers and communications.

This book is about the strategic use of telecommunications in the modern corporation. And yet, it is impossible to discuss this subject, elaborate upon its current progression, and properly place it into a wider context without examining attitudes toward the acceptance of new networking technology by senior managers, middle managers, and professionals in the wake of their encounter with it over the last twenty years. By and large, acceptance has come very slowly. The reasons for this are complex and have much to do with the power of technology (in this case, electronic communication) to shape and transform traditional organizational structures and business modalities in completely unpredictable ways.

In part, the problem is one of perception. So fundamental is the process of electronic communication that it tends to eventually become "invisible" to those who use it. This transparency (to use a term that has a special resonance in the communications manager's lexicon) is rarely examined or viewed as something to be managed, changed, restructured, or otherwise tinkered with. Such processes are often either left to chance by management or delegated to other functions within the organization that have little real capability or corporate clout to alter them. And yet, think for a minute about the very nature of business and commerce: an activity that inherently relies on the communications process as its fundamental *modus operandi*. For example, marketing—that most basic of corporate tasks—is simply the process of promulgating the value of goods and services to potential and actual buyers. Marketing is essentially carried out via communications in a variety of media, including advertising, public relations, direct mail, shows and exhibitions, and so on. Each of these media can be examined to determine whether the communications process employed is best suited to the nature of the message. To choose another random example, the processes of inventory control and accounting represent other basic functions that, once deconstructed, yield a simple enough process: the flow of information about corporate financial transactions to a central organizing authority.

When one begins to look at all corporate functions and procedures through the prism of communications analysis, the results yielded are interesting and often disturbing. This process of analysis invariably reveals how little actual managerial thought and planning is invested in the nature of these communications. The tendency not to look critically at infrastructure suggests the existence of whole new areas of exploration regarding the communication of business information. Thus, in the not-too-distant future, companies ought to begin to pay as much attention to the how and why of business communications and is now paid to the who, what, and where.

Yet this particular—some would say McLuhanesque—perspective on the dynamics of corporate business communication hardly represents an area of widespread or common agreement in management circles. There are many who simply do not view the wide array of communications-related corporate functions in these rather fundamental terms, just as management, even in the fast-track 1990s, continues to struggle with the proper role of computer and communications technology. Part of this struggle by corporate managers to define the role of communications technologies may relate to a perceived threat posed by the new technology to traditional hierarchical business practices that have been comfortably in play for many years. In any event, corporate planners often still view information services (IS) and telecommunications-related functions as mundane elements of infrastructure, rather than active, transforming agents and tools that can be used to provide strategic advantages, both internally and externally.

If management was often slow to recognize and implement these new technologies, then how did IS/telecom managers cope with their part in the evolution of these new systems? Conventional wisdom holds that traditional large-company MIS shops became too bureaucratic, resulting in the inability to develop a real appreciation for the needs of their own internal customers (that is, departmental end users) [2]. Another often-heard criticism is that MIS shops ultimately succeeded in using computer technology not to make significant improvements in basic business operations, but rather to automate what were in many cases poorly designed procedures to begin with. Regardless, it is clear that the dynamic here is organizational in nature. Without mandates from senior management, MIS was not in a position to chart radical new approaches as it began to implement systems companywide. Furthermore, it could be argued that the slow pace at which corporations adopted these powerful new technological tools stemmed from the fact that managerial or organizational structures were simply not ready to adjust themselves to the radically new social vectors that the new technology unleashed.

There are many other factors affecting the degree of success with which computer and communications technologies have been implemented over the last twenty-odd years. Many of these factors are cultural in nature and will be explored in later chapters under the rubric of more specific technologies. Moreover, it bears repeating that there are many corporations that have indeed begun to implement innovative, strategically significant uses of network computing applications and are realizing significant payback in terms of both the cost of the systems themselves and their impact on the corporate "bottom line." These leading-edge applications will be discussed in subsequent chapters, particularly in Chapter 7.

THE STRATEGIC VALUE OF COOPERATION

Geopolitically and economically, the rise of the new world marketplace is one of the most important developments of the 1990s. New trade agreements, the extensive liberalization of trade, and the redrawing of nation-state boundaries and policies have indeed created a "new world order" that affects not only the business activities of multinational corporations, but also the sales and distribution strategies of both large and small companies. Writing in the Atlantic Monthly, Benjamin Barber once described, somewhat humorously, this new global marketplace as "McWorld" [3]. According to Barber, there are four major dynamics of McWorld: a market imperative, a resource imperative, an information technology imperative, and an ecological imperative. All of these, it can be noted, have significant interdependencies. More important for current purposes, though, is the essential role that information technology is deemed to play in this emerging new world order.

Generally, analysts see this new economic landscape as dominated by three major segments: the North American market, constellated around the United States; the Pacific Rim countries, centering on Japan but also including the "four tigers" of East Asia—Singapore, Hong Kong, South Korea, and Taiwan; and Western Europe, focusing on Germany. Much of this new order in commerce and business is predicated on the continuing use of electronic communication in all its forms. Thus, communications technology has now become a major linchpin in the ability of companies to compete and in many cases has become a *sine qua non* for sophisticated new global market players.

There is, however, another side to telecommunications and computer networking, one that in business management circles is not frequently discussed in the endless handwringing sessions over America's loss of competitiveness: its role in the need for cooperative teaming. Fortunately, this important but long-overlooked area is beginning to enjoy some attention as corporations become more willing to adopt new ideas and examine the mainstream management and business practices that are the norm in countries such as Japan.

In this context, a frequently asked question crops up: is America losing market share and dominance in the new global marketplace because of a socioculturally conditioned orientation that discourages cooperative teaming? Invariably in such discussions, cooperative mechanisms and arrangements that Asian countries—especially Japan—have successfully adopted are cited. Increasingly, the answer to this question seems to be in the affirmative. In 1986, a study was undertaken by the Massachusetts Institute of Technology, the ambitious objective of which was to place under objective scrutiny the performance and business practices of a wide array of U.S. vertical-market industries. The industries examined included consumer electronics, automobile manufacturing, steel making, computers, textiles, and chemicals, to name a few. The findings of the study were published in 1989, appearing in the report *Made in America: Regaining the Productive Edge*. Among the most significant problems found in these (and other) industries was *a lack of cooperation throughout all levels and areas of American business* [4].

Problems such as this are structural and, hence, difficult to address and correct. For one thing, they involve sociocultural factors that are deeply embedded in the ways in which managers and workers view the responsibilities, demands, and requirements of the workplace in general. For example, American attitudes toward cooperative teaming have been found by many researchers to be significantly different from approaches that are taken for granted in Japan and other Asian nations. American concepts of business success have always placed a high value on the virtues of individual initiative as the best means of fulfilling both corporate and personal career objectives. In contrast, Asian thinking about such matters tends to be oriented considerably more toward the spectrum of considerations affecting the group, with an attendant emphasis on collaboration and cooperation rather than competition.

Japan, in particular, has been the focus of much managerial soul-searching in terms of contrasting management and employee styles, revolving around the concept of *keiretsu*. This is an enterprise model that relies on significant levels of cooperation between manufacturers, suppliers, and financiers. In the United States, however, changes in traditional approaches are already in evidence. As described in *Business Week*, hundreds of companies in industries as diverse as computers, semiconductors, automobiles, and farm machinery ". . . are revamping their cultures and recasting their investment practices to form cooperative links both vertically, down their supply lines, and horizontally, with universities, research labs, and their peers" [5].

Interestingly, although such concepts are indeed being examined in the business trade press and other corporate forums, the subject of electronic communications as a potential enabling technology to bring about such changes has yet to be more widely acknowledged. Moreover, as issues such as the sharing of information via computer and communications become more widely evaluated, these differing cultural viewpoints will become even more highly contrasted. For American corporations, some interesting choices are looming ahead with respect to how necessary it is to move toward more cooperative teaming approaches in the future. This is where, from a computer networking standpoint, things get interesting, because it's entirely possible that the kinds of new technologies discussed in later chapters—such as EDI, groupware, and workflow computing—will come to represent the necessary vehicle to transcend the cultural determinants that lie at the base of the problem.

NETWORKING AND ORGANIZATIONAL PARADIGM SHIFTS

The forces driving the need for increased cooperation between U.S. corporations and their business partners, customers, and suppliers (as well as those involving new partnerships among elements of federal, state, and local governments) also affect the internal structures of the corporate organization. One of the most fundamental changes here is the so-called "flattening" of organizational structures, whereby communication increases horizontally among various workgroups, a process fostered and facilitated by new desktop-based communications technologies. In many cases, this new emphasis on horizontal communications has created altogether different patterns of intracorporate information flow within companies, patterns that have been well documented in numerous contemporary studies of organizational behavior. Various explorations of this phenomenon, such as Charles Savage's *Fifth-Generation Management*, describe how the patterns of industrial-era thinking, with their emphasis on regimentation and uniformity in group behavior, still lie at the root of modern management practice [6]. Savage says that these patterns must be "unlearned," and an attempt must be made to move away from a mind-set that emphasizes narrow views of job responsibilities and fosters the kind of compartmentalized thinking and activities that bureaucracies are famous for. Interestingly, these new forms of organization, as will be discussed later, mirror and parallel the new expressions of computer networking capability that continue to support their evolution.

The forces thus far described can be viewed as powerful externalities acting upon corporations large and small and effecting major transformations in the way companies do business. These transformations are thus often responses to the vectors and dynamics of the new global marketplace. Computer networking is at the heart of some of these changes and will interactively shape the ultimate outcome of how companies respond to them. Apart from these larger trends and developments, however, there are specific tactical and strategic steps that companies can take to build a sustainable advantage for themselves as they use the computer-based capabilities of time- and distance-insensitivity to extend their markets both geographically and chronologically. These kinds of applications are discussed in detail in Chapter 7, which describes a series of successful applications developed by leading-edge early adopters, many of which are Fortune 1000 companies. Therefore, it's important to draw a distinction between the strategic use of communications in sharpening and optimizing the performance of internal operations (and thus gaining a competitive edge) as opposed to using new computer and communications technologies as application tools in the service of a business plan that specifically calls for the use of such technology. Both of these areas are addressed in subsequent chapters. Chapter 4, for example, deals with optimizing intracorporate communications; Chapters 5 and 6 deal with organizational issues affecting the development of both dimensions. Chapters 3 and 7, on the other hand, maintain a stronger focus on specific strategic applications—specifically, how they are optimally developed and what makes them successful.

The focus in many industry discussions to date has often been on this latter aspect. Indeed, much of the convergence of business theory with communications theory has centered on these kinds of applications. Peter Keen, in his book *Competing in Time*, provided some excellent descriptions of companies that managed to accomplish first-strike marketing coups with new and innovative approaches to the technology. However, Keen's book tends to dwell on big-systems applications, which, I will argue in later chapters, are only a part of the overall picture with respect to the challenge ahead for IS and telecom managers.

Another analyst who has made a significant impact on business theory in the context of strategic communications is Harvard University's Michael Porter. One of Porter's most important contributions to the current body of thinking is the concept of the value chain. Simply defined, the value chain is the sum total of various functions and operations that make up the gestalt of a company's competitive strategy. (Typical components of the value chain are shown in Figure 1.1.) The final objective of these elements is to optimize corporate profit margins, which can be accomplished only when all of them are fully orchestrated in relation to one another. The important point here is that a well-conceived IS/telecom plan has a role to play in just about every one of the strategic building blocks shown in the diagram. In other words, each of these activities requires the process of communications to take place in the routine conduct of daily operations. Accordingly, electronic communications adds value to each of these areas and allows them to be optimized according to their individual functional profiles.

The value chain, then, reinforces the previous distinction between optimizing internal operations and developing strategic applications. For the purposes of assessing impact on the value chain, we can call the former *strategic optimization*

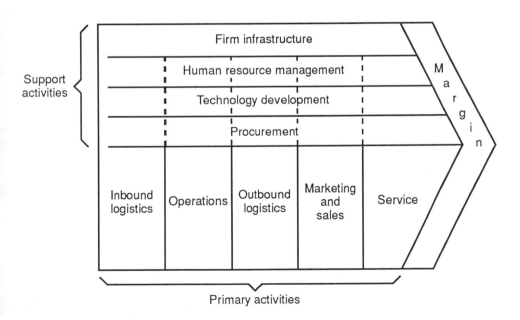

Figure 1.1 Porter's value chain. (From Elbert, *Private Telecommunications Networks*, Artech House, Norwood, MA, p. 388.)

and the latter *strategic implementation*. In the case of the former, the idea is to take all of the standard departmental functions within the corporation (human resources, accounting, purchasing, and so on) and apply state-of-the-art computer and communications to the task of optimizing the procedures and practices of each. In the case of the latter, these same technologies are actually used as the principal vehicle for a company's core business activities, particularly with respect to the constellation of its relationships with customers, suppliers. and partners. It is, of course, in this latter area that the critical functions of sales and distribution come into play.

U.S. INDUSTRIAL POLICY AND NETWORK INFRASTRUCTURE

The strategic use of telecommunications can be discussed on many levels. Much of the treatment throughout this book centers on how individual corporations can develop strategic applications using combinations of their existing private network capabilities and those inherent in the evolving data transport infrastructure of the public network. However, strategic networking can also be approached at the macro level in the context of a subject that is fast becoming the focus of attention in business and government circles: the question of whether the United States should adopt a formal industrial policy.

Discussions of industrial policy inevitably involve political and public-policy considerations, such as the proper role of government in stimulating leading-edge technologies that will be turned over to private industry for longer-term development. In some quarters, the very mention of the subject of industrial policy raises less-than-enthusiastic reactions from staunch private-sector advocates. However, rising concerns over the competitiveness question discussed earlier have also prompted many policymakers to ask why the United States is virtually the only OECD country that doesn't have some kind of cohesive approach to fostering business via cooperative partnerships between government, corporate, and academic entities.

In the discussions swirling around this issue, one item that has been put on the agenda is the subject of network infrastructure, in the expectation that it might help foster cooperation among these parties, thereby accelerating the process of research and development investment. Proponents of the NREN, for example, have argued that this network could be used as a major linchpin in U.S. industrial policy by providing a forum for information exchange vital to the seeding of new technologies. (The NREN, for all intents and purposes, can be described as the next generation of the U.S.-based portion of the Internet.) In theory, the NREN/ Internet could indeed be used to create a cooperative communications platform that could tie together new constellations of partners in business, academia, and government. Furthermore, there is the background issue concerning the role of the defense industry in the new information economy (keeping in mind the network's origins in this sector): how its considerable resources can be productively channeled into improving the quality of life and how technology transfer from the defense to civilian sectors can develop new economic synergies for the commercial marketplace.

In attempting to look objectively at this issue, it may be helpful to draw distinctions between an enabling role on the part of government versus an ongoing "hand on the tiller." Many precedents for the former exist throughout the course of U.S. history. For example, the federal government has often played a role in developing various kinds of infrastructure, including railway systems and the vast connecting web known as the U.S. interstate highway system. Both of these efforts were widely judged to have been successful; however, in both cases, the government provided seed money, along with economic impetus and political will for new industries to grow and expand, and then removed itself from the picture. In much the same way, various business and industry proponents argue that the federal government should do the same for yet another type of infrastructure: communications networks. They also argue that the NREN is the logical vehicle for developing this infrastructure.

Policy issues concerning the NREN/Internet are extremely complex and, accordingly, will be outlined but not elaborated upon in this and later chapters. However, if new cooperative and collaborative approaches are to be adopted industrywide, then a national utility network (or at least regional segments of the existing Internet) could provide a platform for doing so. Furthermore, there have been proposals put forth to give the Defense Research Projects Agency (DARPA) a major role in industrial policy development, in line with its traditional role in providing initial funding and resources for defense projects. The interesting tie-in here is that DARPA was the federal agency first responsible for developing the Internet some twenty years ago when it was known as the ARPANET. In any event, suffice it to say that there is a plethora of public-policy issues to be dealt with in this newly emerging arena, most of which will not be resolved overnight. In working toward this goal, however, policymakers will have to consider the role of corporate private networks as they already exist, deployed throughout major corporations; the emergence of new data transport capabilities in the public switched telephone network (PSTN); and how, if the NREN is chosen to be the platform for these new cooperative and strategic partnerships, its capability can be extended to other enterprises, both large and small, with equitable access for all subscribers. This will be especially important, given the new realities of downsized and restructured corporation organizations, as described in detail in Chapter 3 [7].

NOTES AND REFERENCES

[1] For more on this subject, see Thomas S. Valovic, "Let's Get Computer Networking on the National Competitiveness Agenda," *Telecommunications*, April 1992, 4.

[2] A note on the use of terminology throughout this book: Generally, for the sake of simplification, I have tried to use MIS as a historical term, referring to the traditional "glass-house" mainframe environment circa 1980–90. The term IS, by contrast, is generally used to refer to a revised or reinvented sense of this function as is generally now used throughout the industry.

[3] Benjamin Barber, "McWorld vs. Jihad," *Atlantic Monthly*, March 1992.

[4] For more information on the MIT study, see Susanna Opper and Henry Fersko-Weiss, *Technology for Teams*, Van Nostrand Reinhold, New York, 1992.

[5] "Learning from Japan," *Business Week*, 27 January 1992, 52.

[6] Charles Savage, *Fifth-Generation Management*, Digital Press, Bedford, MA, 1990.

[7] For more on this subject see the following articles:

Vinton G. Cerf, "Another Reading of the NREN Legislation," *Telecommunications*, November 1991, 29–30.

Jay Habegger, "Why is the NREN Proposal So Complicated?" *Telecommunications*, November 1991, 21–26.

Thomas S. Valovic, "ANS and Internet Commercialization," *Telecommunications*, July 1991, 4.

———, "A Giant Step Towards Internet Commercialization," *Telecommunications*, June 1991, 7.

———, "Is NREN an 'End-Run' Around MFJ Policy Gridlock?" *Telecommunications*, January 1991, 9.

———, "The NREN Enigma: A New National Network?" *Telecommunications*, January 1991, 13–14.

Chapter 2
Technology and Infrastructure

In the last fifteen years a revolution in technology has transformed the way in which network engineers think, plan, and build. The network has changed from something like a pyramid to something more like a geodesic dome. This architectural transition from Cheops to Buckminster Fuller has occurred with breathtaking speed.

—Peter Huber

Telecommunications technology is a moving target. As such, descriptions of trends and patterns in its evolution often seem to become outdated as soon as they're examined. However, the trajectory of technological growth and development in the computer and communications arena is distinct, and larger patterns are observed that exhibit reasonable consistency over time. Downsizing and decentralization, for example, are trends that fall into this category. This chapter will look at emerging technologies in the light of their relationship with major industry patterns of development and, more specifically, how they—and the infrastructure they support—will affect the future shape of corporate networks and their strategic use. (Nontechnical readers who are unfamiliar with the technology described here should not feel constrained to read through this chapter and may go on to Chapter 3.)

The explosion of new computer and communications technology that took place in the early 1980s empowered corporations and individuals in ways that were unique with respect to traditional business methods and practices. Although much of this technology had already been quietly incubating in the technological hothouse that was AT&T's predivestiture Bell Laboratories, it took the deregulation of the industry, along with changes engendered by divestiture, to uncork the Bell Labs genie and allow the dissemination of networking-based intelligence outside the carefully protected boundaries of the public network itself. This phenomenon was aptly described by Peter Huber in the landmark report on the state of U.S. tele-communications, *The Geodesic Network*:

> As the vital center of equipment markets has moved away from the central office out to private premises, control over the transmission-switching tradeoff has moved from carrier into private hands. And these private hands now bid fair to wring their mother's neck. If every LEC shut down tomorrow, the market for PBXs, computers, and data processing services, and a vast array of electronic storage and retrieval systems that interconnect with the public network, would only pause briefly and then boom. [1]

And boom it did, with the proliferation of thousands of new types of private networking devices, including multiplexers, modems, data encryption equipment, PBXs, protocol analyzers, PADs, access devices, and so on. An idea of the extent and scope of this rapidly expanding marketplace can be gleaned from the number of new ventures and products created during the second half of the 1980s. According to one respected tabulator of such statistics, cited in a report compiled by the North American Telecommunications Association, in 1975 there were approximately 850 products in the marketplace [2]. In 1985, that number had jumped to 1150; by 1988, it had skyrocketed to 1349. The report also noted that the number of manufacturing establishments producing communications equipment has jumped almost 8% per year since divestiture. Seen in this light, divestiture/deregulation should be viewed positively. Put another way, had the requirements for modern corporate high-speed voice and data communications emerged as vigorously as they did in a scenario that did not include divestiture, then it is highly questionable whether the traditional Bell system could have been responsive enough to meet those requirements, especially in the case of the high-end needs of the Fortune 1000.

A major precipitating factor in these developments was the rise of data communications in the early 1980s as, increasingly, computers were linked either to users (via terminals) or to each other. If data communications took place on an organization's premises, they were typically referred to as local-area network (LAN) communications. If extended geographical areas (that is, different cities) were involved, then the use of a wide-area network (WAN) was typically involved, requiring the use of the analog-based public switched telephone network (PSTN). In fact, during the last twenty years or so, an entire industry has been built around a simple anomaly of the PSTN: the inability of voice-grade analog lines to accommodate the digital signals associated with computers and data communications. The simple need to convert digital data into analog signals that could traverse telephony networks created the demand for a device called a modem (modulator-demodulator) that made the necessary conversion. As a result, a multibillion-dollar industry took root, providing modems, multiplexers, and the other equipment geared to optimize the PSTN for the more sophisticated and reliability-sensitive demands of data transmission.

THE FUTURE OF THE PUBLIC SWITCHED NETWORK

It's impossible to appreciate how corporate networking requirements have evolved over the years without first looking at the major technological driver: the basic use and applications of the computer. Computer technology can generally be described as being reasonably mature, at least in terms of what might be called the first phases of its development, from UNIVAC behemoths to the most intelligent and downsized palmtop. Truly innovative voice and data communications technology, however, got a much later start (at least from the perspectives of those vendors and providers that functioned outside of the traditional Bell system) and really didn't begin to evolve in a truly dynamic fashion until the mid 1980s. Thus, it seems safe to say that this exciting revolution in communications technology has only just begun. In fact, certain trends and patterns that were part and parcel of the technological trajectory of the computer industry are now beginning to manifest themselves in the communications realm.

How are these patterns best described? In general, there are several important dimensions to technological growth, including innovation, standardization and commoditization, interoperability, and miniaturization and downsizing. The computer industry has experienced each of these throughout its various cycles. For example, the personal computer (PC) revolution spawned a great proliferation of hardware and software alternatives. In this Darwinian, free-market survival-of-the-fittest technological explosion, the best and most popular systems usually survived in the long run. Later, however, in a more mature market phase, many companies, after a rash of renegade PC departmental purchasing, found themselves saddled with a corporatewide crazy quilt of widely disparate and incompatible hardware, software, operating systems, and peripheral equipment. Each of these systems had different command sets, communications requirements, graphical user interfaces, and operational needs.

When this explosion of competing and incompatible systems has finished its cycle, a different market dynamic takes over, characterized by the attempt to reconcile fundamental differences between vendors and the basic systems that underlie their products. This is the process of standardization. However, even the pursuit of standards creates its own set of problems, and, in many cases, such efforts often serve only to create more confusion, because manufacturers are notorious for subverting the standards process simply by the sheer drive toward competitive positioning. Thus, when AT&T and Sun Microsystems supported the concept of UNIX as the principal standard for open systems, there was some hope in the industry that it would be a rallying point for other computer vendors. Instead, however, the move was viewed as a declaration of war, not peace, and a group of competing vendors (such as IBM, DEC, and Hewlett-Packard) moved to establish the Open Software Foundation (OSF).

Another observable phase in the growth path of the computer industry has to do with product commoditization and downsizing. As new functions and capabilities are required, they are often developed as stand-alone products. As these products become accepted and more widely used, vendors find that, thanks to the benefits of miniaturization, they are able at some point to incorporate that product's function into another product. In other words, one hardware platform comes to subsume the capabilities of another smaller one. The PC market offers several good examples of this phenomenon. The first is the development of the internal modem. When modems were first developed, they existed as stand-alone products. Eventually, modem functionality could be moved onto a single chip, and the chip could be embedded relatively easily into the PC motherboard. Thus many PCs on the market today—including laptops—have not only internal modems, but also LAN attachment capability, and are labeled "communications-ready." Similarly, fax machines are still in the stand-alone phase of development, but increasingly fax functions are being incorporated into PC hardware in the same way.

Another interesting trend in computing is the relationship of hardware to software. As the market matures, product designers develop a better understanding about which processing tasks should be hardware-resident and which are best accommodated by software. Increasingly, vendors, in order to stay competitive, are designing their systems so that subsequent performance upgrades can be accomplished with a simple software upgrade, done by floppy disk insertion or downline loading via modem or the Internet. Older devices, however, are limited to being field-upgradable, in which case a technician must actually make an on-site visit to replace a PROM chip, a process involving additional time, cost, and effort for both vendor and customer. Another aspect of this trend—what I once described in a *Computerworld* article as "etherealization"—is that products and product functionality continue to become more software-driven. As hardware becomes more standardized and computer platforms actually revert to being "general-purpose" in function, software developed by vendors or third-party providers increasingly affords the primary share of product differentiation and value-added functionality.

Extrapolating the trends driving the computer industry to the communications realm proves to be a challenging exercise. However, it is without question a useful exercise for developing a wider appreciation of future directions in the communications industry. These same trends—innovation, miniaturization, commoditization, and etherealization—are equally applicable in assessing the impacts of communications products on strategic use and applications. However, as mentioned earlier, there is a distinct lag time associated with the degree to which some of these phenomena have taken hold in telecommunications. In many cases, therefore, we are only at the beginning of an anticipated "S-curve" in their evolution.

In the realm of commoditization, for example, many observers envision a public network that adheres essentially to the technical model of a general-purpose platform upon which specific applications are implemented via software development. This is in sharp contrast to the massive hardware changeovers and long lead times that changes in the public network have traditionally required. A specific conceptual model for how this might actually work was advanced by the International Communications Association (ICA), a prominent user group. The ICA, in fact, suggested that public network providers—primarily the Regional Bell Operating Companies (RBOCs)—should more explicitly adopt the model of the hardware platform as the basis for the same kind of software-differentiated functionality that the computer industry has already developed. An ICA report entitled *New Connections for the 1990s*, explained it this way:

> When the PC model is extended to telecommunications, the local network's primary role becomes that of the basic applications platform—providing transmission, switching, and other services that are most efficiently or exclusively provided through the ubiquitous local exchange network. In some instances, this platform will consist of unbundled basic switching and transport; in other instances, new intelligent functions will be married with the telephone company's own basic switching. The local network and its operating system should support other intelligent applications created by third parties. These *applications developers*, including but not limited to the telephone company, have a separate role. They focus on inventing new features that utilize the evolving capabilities of the local network, while *applications users* translate those features into new services and products that are of direct practical use to the *end user*. [3]

In this view of the future public network, the network is fully transparent to applications, and independent service providers might then wish to write specific software-based application packages for the network, much as they did for PCs.

Conceptualizing this kind of approach is one thing; actually seeing it translated into technical reality is quite another. In an ideal world, the intelligent network nodes of the PSTN would indeed be open programmable platforms in which users could enjoy far more direct control over changes in service delivery—the kind of control they've been accustomed to on private networks. In point of fact, this kind of progress has been taking place, however gradually. Companies such as Tandem and DEC that sell minicomputer-based hardware platforms have been working to conform to long-range RBOC objectives in fulfillment of Bellcore's master blueprint, the advanced intelligent network (AIN) concept. Such approaches will allow both local exchange and long-distance public network providers to roll out flexible service arrangements and reconfigurations far more quickly than was possible previously. Such responsiveness to ever-changing customer requirements in the new competitive environment is a critical dependency in the efforts by public network providers to keep revenue-generating corporate customers from relying increasingly on private network solutions that bypass the public network.

Another important area that involves providing more flexible functionality in the public network is the concept of open network architecture (ONA). For many years, the FCC has been working to develop ONA as a regulatory model for enhanced service providers (ESPs) who wish to provide their services under the auspices of the public network. The driving principle is that various public network functions could be made available to these independent, entrepreneurially driven network providers on an unbundled basis. In order to do this, however, the existing array of public network services needs to be broken down into its basic building blocks. In the language of the ONA provisions, these are known as basic service elements (BSEs). In theory, these units of functionality are then made available to individual ESPs at the same cost that the RBOC "pays" for it.

Another view of how the public network might evolve into a flexible vehicle for independently generated applications has been articulated by Mitchell Kapor, founder of the Lotus Corporation and now head of a lobbying organization called the Electronic Frontier Foundation:

> The most important contribution of the PC field is not a product, but an idea. It is the idea that a good computer system is simply a platform upon which other parties can exercise their ingenuity to build great applications. . . . In short, it was the existence of open platforms for innovation, such as the Apple II and the IBM PC, which catalyzed the development of vast amounts of software necessary to the process of market-mediated innovation. Today, with the desktop PC, common-place in business and the home, it's important to remember the basic dynamic by which this PC revolution occurred. Just as the desktop personal computer represented the revolutionary platform for innovation of the 1980s, it is my belief that ubiquitous digital communications media, such as are enabled by the Integrated Services Digital Network (ISDN), represent the hope of the 1990s. With the proper ISDN platform, we can have another generation of explosive growth of services, led by a generation of information entrepreneurs. Today these information entrepreneurs enjoy a marginal existence in the largely noncommercial world of bulletin boards and on the national research and education network called the Internet. Give them a commercial information infrastructure which can reach large numbers of people inexpensively, and I believe we will all be truly amazed at the results. [4]

It's worth noting that there are many in the telecommunications industry who might disagree with Kapor's assessment of ISDN as the ideal PSTN communications platform. Nevertheless, the point stands that the public network needs to develop standardized, commoditized platforms that can serve as the basis for the kind of entrepreneurship that drove the computer industry so successfully and is the basis for Kapor's enthusiasm.

THE CONCEPT OF THE NETWORK UTILITY

An important consideration regarding infrastructure is the emerging concept of the corporate utility network. Much closer to ideal than reality, the concept presupposes a corporate infrastructure that transparently handles all media—voice, data, and imaging—and allows any user to take advantage of instantaneous interconnection. The concept is analogous to the way in which electric power is distributed via a nationally standardized distribution grid. In this environment, all electrical devices and appliances within the United States are required to conform to codes that allow them to to be engaged with the grid via any electrical outlet regardless of location. By contrast, many electrical devices in the United States are unusable without adapters in Europe or other parts of the world. The idea of a corporate network utility is much the same. Ideally, in any corporate location, individual users would be able to "plug and play" a wide variety of devices—be it a laser printer, Macintosh computer, fax machine, 3270 terminal, or laptop computer. This is of course the goal of standardization and interoperability.

Needless to say, most corporations are a long way from that goal, and the various methodologies devised to allow this kind of infrastructure to be implemented are still evolving. One of the earliest of these was the concept of ISDN. In the 1980s, when AT&T and the RBOCs were promoting ISDN via advertising campaigns, they attempted to make the case that ISDN would usher in a new world of what they called universal information services (UIS). The technical basis for UIS was that a seamless, standardized ISDN network throughout the United States would carry all forms of voice, data, and video via a variety of devices. What this vision failed to take into account, however, were the rapidly evolving differences between the needs of high-end corporate users and the kind of networks required by small- and medium-sized corporations and the general public. Many of these differences were purely bandwidth-related, so the gap widened in the 1980s, with only the prospect of fiber-based delivery of TV signals acting as an influence to close that gap. Nor did that view of the public network accommodate other differences evolving in the university community, where powerful Crays dominated the landscape, and bandwidth-hungry workstations were being deployed. ISDN turned out to be not nearly adaptable enough to be all things to all users.

So if ISDN was not going to provide the corporate utility network infrastructure, then what would? One of ISDN's major monkey wrenches turned out to be the PC. The PC, an inherently and radically high-bandwidth device, and its impetus to spawn islands of LANs, became the force for pushing bandwidth needs of corporations well beyond the meager 64 kbps offered by narrowband ISDN. Yet, as these new LANs evolved, they also began to quietly militate against the ISDN concept by creating sets of networks and subnetworks quite distinct from both traditional IBM SNA and voice telephony networks that were standard items in most corporations. By the early 1990s, tying all of these into any kind of seamless and neat infrastructure began to look like an impossible task. Accordingly, one of the biggest challenges for network planners in the early 1990s was the task of collapsing so-called "parallel networks" (that is, LANs and router-based networks) that were separate from traditional hierarchically based SNA networks and that insinuated themselves into the corporate environment at the departmental level. Thus, although the concept of the utility network is an important goal to work toward, corporations are still a long way from achieving it.

THE INTERNET: A NEW PUBLIC NETWORK INFRASTRUCTURE?

In terms of public network infrastructure, another important trend is the emergence of commercialized Internet services, or what one knowledgeable observer has described as "one of the most important paradigm shifts of the 1990s" [5]. This subject will be treated more thoroughly in Chapter 4; for the moment, suffice it to say that the significance of this development lies in the fact that the Internet can be viewed in one sense as the world's largest public data network. However, up until recently this huge, globally deployed "network of networks" has been operating under acceptable use restrictions that address the objective of preserving its original purpose: research and education. This is changing rapidly as portions of the network are becoming both privatized and commercialized.

As the Internet becomes more available to enterprises outside of this traditional purview, it will further develop its potential to become a major new infrastructure for data traffic—and perhaps voice, video, and multimedia transmission as well—extending to a wide range of constituencies, including corporate, commercial, and residential user groups. In that sense, it is on the way to becoming a public network, in the traditional sense that the term is used, especially as its U.S. portion evolves into the NREN. This makes the birth of the commercialized Internet an extremely significant event, possibly even akin to the development of the first nationwide AT&T voice telephone network, but with the additional significance of being global in reach, multimedia-capable, and based on a computer and communications standard that has already received wide acceptance among corporate users [6]. Interestingly, such a reality is strikingly similar to the hopes put forth by early proponents of OSI-based ISDN. However, although ISDN will eventually become fully deployed, it will remain oriented toward the lower layers of the OSI model as a more transport-oriented standard.

Much of the value of the Internet is related to its roots in the transmission control protocol/Internet protocol (TCP/IP) protocol stack. Whereas TCP operates at the equivalent of layer 4 of the OSI model, and IP at layer 3, TCP/IP is often used to refer to an extended architectural model that has developed around those two layers (see Figure 2.1). TCP/IP is the workhorse of interoperability and has proved itself where it counts: in user networks. The protocol was developed in the

1960s by DARPA, in conjunction with the ARPANET and other networks that were the predecessors of the Internet. It was then formalized by the Department of Defense in 1980.

Despite its origins in government, research, and academic environments, TCP/IP eventually gained increasing commercial acceptance, achieving the status of one of the industry's most important standards. As an internetworking protocol, it is comparable to the OSI model being fostered by ISO. However, because of ideological foot-dragging on the part of the vendor community, resulting in a lack of OSI-based products in the commercial marketplace, OSI lost considerable momentum in the United States during the early 1990s, even though support for it remained strong in Europe. TCP/IP moved in to fill the void, because it was already being used successfully to solve interoperability problems. This trend was epitomized by a major statement from IBM to the effect that the company would in the future place support for TCP/IP products on equal footing with OSI products. Figures made available in 1992 from market researcher IDC indicate that as many as 60% of multiuser systems already have some form of TCP/IP connectivity. Another important factor affecting TCP/IP's momentum is the fact that UNIX is gaining in popularity, because TCP/IP capability is inherent in the UNIX kernel.

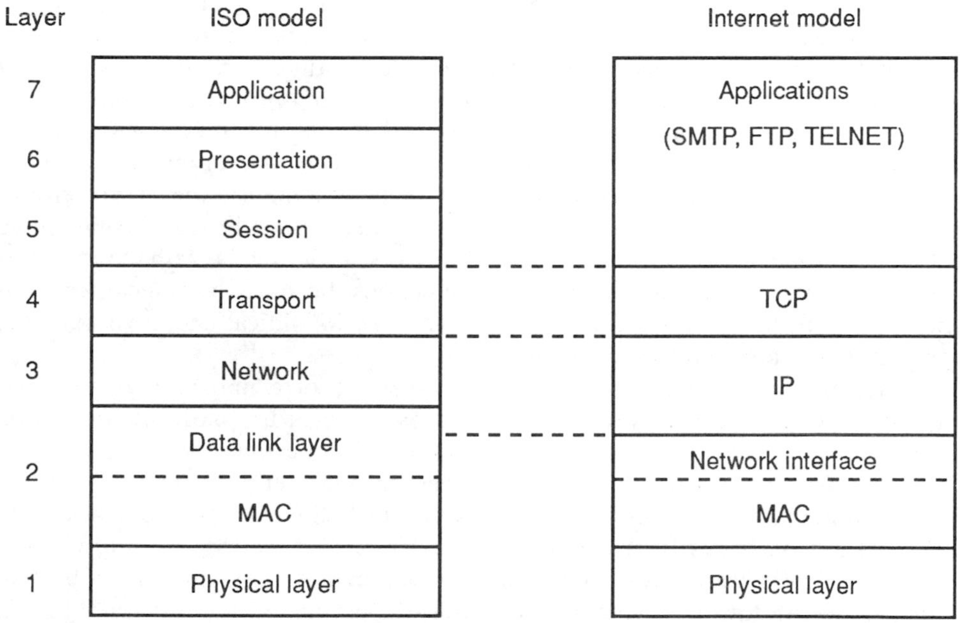

Layer	ISO model		Internet model
7	Application		Applications
6	Presentation		(SMTP, FTP, TELNET)
5	Session		
4	Transport		TCP
3	Network		IP
2	Data link layer		Network interface
	MAC		MAC
1	Physical layer		Physical layer

Figure 2.1 A comparison of the OSI and TCP/IP protocol stacks. (Source: *Telecommunications*, April 1991.)

When looking at the protocols, it's interesting to note a major difference in how they've developed. With OSI, the standard has been developed on a top-down basis, evolving from the "mindware" of its planners to a conceptual standard and then to working product. In the case of TCP/IP, however, the process is somewhat reversed. Proponents of various TCP/IP-related standards tend to be actual users, rather than pure planners. Hence, the standard gets hammered out at the operational rather than the conceptual level.

Because UNIX was generally found to be more prevalent in the research and academic community, so was TCP/IP. To the extent that UNIX is beginning to move into the commercial marketplace, so is TCP/IP. What's even more interesting is that this shift has happened at a time when multiprotocol LAN islands were being developed, and sufficient numbers of OSI products were not available to provide the kind of internetworking capability needed to support connectivity. Thus, the early 1990s witnessed the following situation emerging: increasing use of TCP/IP in the commercial sector, increased need for LAN interconnection, an only partially developed public network infrastructure to handle these needs (that is, SMDS, frame relay, and B-ISDN), and the commercialization of the TCP/IP-based Internet in full bore.

There is much speculation as to where the trend toward commercialized Internet services will lead and whether, in conjunction with the NREN bill passed by Congress, the Internet will become the basis for a nationwide and worldwide data network that will become the analog of the existing voice network. *At least in theory, with the proper oversight by those responsible for its development, the Internet could become a worldwide data networking utility that multinational corporations and enterprises webs could tap into, regardless of geographical location.* In other words, it could at sometime in the 1990s become a principal vehicle for global enterprise networking in support of commerce and, hence, a competitive platform for the development of goods and services. It also might, with the emphasis of the original research and education connections, become the vehicle for commercial involvement in solving a multiplicity of sociopolitical and environmental problems that need not be catalogued here.

Before such seemingly grandiose scenarios devolve into reality, however, many technical, regulatory, and economic issues regarding network utilization remain to be addressed by telecommunications policymakers; Congress; the National Science Foundation, which oversees the Internet's backbone operation; and the Internet's governing bodies themselves, including the the Internet Society, the Internet Activities Board (IAB), and the Federal Network Advisory Council (FNAC). From a marketing and technical development standpoint, other important impacts on the future of commercialization will be made by the major Internet providers, including Advanced Network and Services (ANS) and the member companies of the Commercial Internet Exchange (CIX), such as Performance Systems International (PSI), UUNET Technologies, CERFnet, and Sprint.

THE VIEW FROM MIS

Thus far, we have presented several views regarding the concept of infrastructure. Needless to say, such considerations are vital for companies that need to engage in long-term planning for their network requirements. What the public network or the commercialized Internet will look like in five years will have a significant bearing on which capabilities need to be developed in-house and which can be procured more economically from various public network providers. Not surprisingly, in the context of the MIS environment, computer professionals have their own concept of how a corporate utility for computers and communications should be developed. Also not surprisingly, such a concept has tended to be referred to not as a corporate network utility, but rather as an information utility. Furthermore, there have been substantive differences between these approaches and those developed by tele-communications planners.

Tangled up with the concept of the information utility are several other somewhat controversial management issues related to corporate networking, including the role of the chief information officer (CIO) in the senior management organization and the use of IS for competitive advantage. For example, after some early experiments with using IS for strategic advantage, some industry experts came to rethink the nature of its ultimate role within the corporation. The gist of this new thinking was that IS departments should become not the central organizing principle behind the use of strategic systems, but rather facilitators—service organizations, in essence—dedicated to placing user-friendly, highly networkable computing power into the hands of end users. One of the principal proponents of this decentralized concept was Max Hopper, a high-profile IS executive who worked for American Airlines and was largely responsible for developing their strategically oriented (and highly successful) SABRE system.

There are indeed powerful arguments for this notion. Some of them relate to the fact that developing strategic systems on a centralized basis—a big-systems approach—is simply too ambitious an undertaking for any group of IS/telecom professionals to take on. Such efforts, for example, require the IS staff to develop an in-depth understanding of each unique species of business activity within the company, on a department-by-department basis. Instead, the new thinking queried: why not put the computer tools in the hands of specific groups of professionals and let them shape those strategic applications on their own terms, and in ways meaningful to them? As one observer put it:

> The information utility is a metaphor for the way companies design their technology infrastructures. The focus . . . should no longer be on designing large-scale strategic systems that give a firm a competitive advantage; instead, companies should take advantage of cheaper, more powerful computers, easier to use software, simplified connectivity, and other

off-the-shelf technologies to extend computing power to all workers, enabling them to put their analytical skills to work to create their own strategic systems. This idea, particularly coming from Hopper, may strike some as disingenuous. After all, he owes his high profile in part to American Airlines Inc.'s computerized reservations system, SABRE, which has epitomized strategic systems for more than two decades. But, argues Hopper, systems such as SABRE, which require millions of man-hours and billions of dollars to create, are the sort of technological home runs that companies can no longer afford to build their game plans around. They are too expensive, take too long to build, and—his decisive point—they provide only short term advantages because technology now permits competitors to quickly copy such systems. Instead, by building infrastructure modeled on the information utility concept, IS can prepare its team for a fast pace game of singles and doubles, often hit by end users themselves. No longer a long-ball hitter, MIS will assume the role of coach and trainer. [7]

One of the most important dimensions of this new attitude is the simple recognition on the part of more forward-thinking IS managers that empowering individual users with computer and networking technology is at the critical core of redefined approaches to the strategic use of corporate networks. As such, this trend, along with that of business reengineering, must be considered among the most important developments to emerge in the 1990s. What it means for corporate users is that, in what some have called the "post-IBM area," the glass house will be replaced by the "glass web," creating new distributed forms of fiber-optic-based high-bandwidth network computing. Furthermore, these new approaches will foster the advent of a grand nexus of pathways for information flow, designed to connect users on a true need-to-know basis, rather than on the basis of constrictive industrial-era management techniques.

Yet even as businesses develop a keener appreciation for the mistakes of the past and exciting possibilities that lie ahead, most experts agree that we are a long way from this ideal. For one thing, companies still have a lot to learn about how to optimally organize their IS/telecom functions in order to bring them in line with strategic objectives. This is to be expected, and trial and error will continue to be the norm for many years to come. The author of the article just quoted, for example, notes at least the possibility that this new line of thinking, as proposed by Hopper, could actually serve to undermine the very concept of having a chief information officer at the helm of the information utility. Others will no doubt argue that most companies will continue to require executive-level representation for this important business function. What is threatened, however, is the notion that the CIO can and must be responsible for everything that takes place on his or her watch pertaining to the use of corporate computer resources. And that notion needs to be bolstered by the affirmation that network computing power positioned—and wisely encouraged—at the level of the individual user represents the best competitive advantage any company could have.

But what will users themselves do with all that computing power? This, in fact, will be the central IS and strategic telecom challenge for the 1990s. As the commoditization of computers and new data-oriented transmission technology continues, the emphasis will begin to shift more and more to applications. Furthermore, these applications will increasingly be developed by end users themselves, under the aegis of a reinvented MIS department acting in the capacity of facilitator, rather than purveyor of arcane technical wizardry. In order to further understand these changes, it might be helpful to take a closer look at what market research has turned up with respect to likely future trends affecting computing and telecommunications infrastructures. These trends will be the real influencers in determining the design and capabilities of future corporate information and networking utilities.

THROWING STONES AT THE GLASS HOUSE

One of the most important paradigm shifts in the evolution of computers and communications in the workplace is related to how their collective role has changed in the perception of senior management. Interestingly, the toppling of hierarchical computing structures typified by the IBM mainframe-to-terminal model has paralleled the flattening of organizational structures, a trend that will be explored more thoroughly in Chapter 3. Along with these changes have come fundamentally new and different views of the best way to utilize and optimize computers and communications capability from an organizational standpoint.

Under the SNA-based architecture of the IBM mainframe, data resided at a centralized facility and, according to the tenets of a kind of trickle-down theory, was meted out in selected doses to those who had the need as well as the official authorization to use it. This approach had both advantages and disadvantages. One advantage was that the quality of information was assured via tight controls over the MIS facilities presiding over the data-collection process. Another benefit was that the problems of security, data integrity, and unauthorized access were kept to a minimum. The major disadvantage was that the data itself, in such a rigidly structured system, could not be easily used (or, more accurately, reused) by end users. Data could be viewed (and this most often took the form of a cumbersome computer printout), but it couldn't, in general, be captured, moved to other documents, manipulated, exported to other systems, converted to other media, and so on. In other words, it wasn't operationally malleable.

The advent of the PC, in conjunction with the rapid rise of LANs, would prove to have a major impact on this traditional hierarchical structure. It placed heavy pressures on these types of networks to adapt to new computing environments, steadily evolving at the grass-roots level. PCs and LANs shifted control of both computing and networking away from traditional MIS and telecom managers and into the hands of individual users at the departmental level. Users, in turn, liberated from the technical and bureaucratic intricacies of the MIS department, wasted no time in exploring these new capabilities and their newfound freedom. Whereas some MIS managers perceived these trends as negative, for others this radical shift in the organizational structure of computing represented an *aggiornamento* for MIS gatekeepers who were tired of having stones thrown at their glass houses.

But what was destined to replace this world of hierarchical computing that had become the very core of MIS operations across a wide spectrum of industries? In a sense, a more flexible but distributed microcosmic version of the mainframe model itself: client-server computing. In the client-server model, a centralized facility for data-base residency still exists. However, that facility is based at departmental or other suborganizational levels. Hence, the new model that began to emerge was both centralized and distributed—in essence, dispersing the "one company, one mainframe" model such that, due to technical advances in computing power and storage capabilities, companies could have the equivalent of a mainframe in every department. As workstations continue to be deployed beyond their originally narrow uses in scientific and research environments, end users at many different levels and functions in the corporate organization are beginning to enjoy the advantages and capabilities of having a "mainframe at the desktop."

These trends and events eventually coalesced into an emerging paradigm for computing in the 1990s: the desktop would become the computing "cockpit," a window on a wide array of virtual realms and capabilities available throughout the network. With the new center forming at the periphery, each end user could become his or her own center of critical information flow, interacting dynamically with other users throughout the network. This particular notion of user empowerment— and the development of strategic applications made possible by it—will continue to be one of the most important trends in computer and communications technologies throughout the 1990s. Now that some of the basic technology is moving into place, further cultural and organizational changes will be needed to reinforce those changes. In addition, new types of collaborative tools, such as groupware, will also play a key role in helping to create new workflow patterns and changing the "mindware" of users. Thus, technology begets cultural change, which in turn begets technology, and so the cycle continues.

According to one knowledgeable source, at least three basic computing models are likely to emerge in the 1990s: centralized, decentralized, and hybrid.

> The centralized model, which looks back to the '80s, concentrates most of the processing power in a few key locations. It provides economy of scale but often suffers from poor response to remote users and requires

many structured and cumbersome procedures to work efficiently. The decentralized model provides local computing power at the user site and is highly responsive to changing user needs. But it is often more expensive, due to the duplication of equipment and the resultant possible underutilization of resources. It makes sense when processor types are strictly limited by space or cost considerations. The hybrid model is the direction in which most organizations are moving. It provides an architecture that offers both local and remote resources to users via networking. It is the most effective method of giving users access to the types of system resources that best fit their problem-solving needs. In the hybrid model, the system focus is of the user at his workstation, where it should be, instead of on the various supporting data processing facilities. The workstation becomes the window into the available system resources, and the problem the user is attempting to solve can be directed to one or many different specialized system elements simultaneously for resolution in a transparent (or at least translucent) manner. [8]

In this scheme, a key supporting technology will be client-server-based distributed processing, as just described.

NETWORKING PARADIGMS SHIFTS

Managers with responsibilities in IS, departmental network computing, and strategic communications will increasingly have to adapt to these new realities, as will front-line departmental managers. They should expect that during the 1990s, businesses will be in what might be described as a continuous experimental mode with respect to computer networking. *Furthermore, the focus of competitive advantage will shift significantly in favor of those companies and professionals who can learn to leverage these important corporate resources quickly and effectively.* For some companies, slow to recognize the new landscape despite the manifold warning signs, it may already be too late. Not only will this new environment favor the smaller, more downsized operation that can put such principles into play quickly, it will also favor companies with the requisite savvy to use existing information networks to promote and market their new services and products.

The fact is that, for all the changes in computers and communications that have taken place over the last twenty years, we are now only at the lower end of an enormous logarithmic growth curve yet to come. These changes (and others perhaps even more radical) will be driven both internally and externally. They will be driven internally by new desktop capabilities that will be available on a commodity basis—new, downsized, and easily deployable technology such as PCN, VSATs, and routers, which will allow users to attach to a dynamic array of high-bandwidth network facilities virtually instantaneously [9]. They will be driven externally by the need for companies to run their business more cost-effectively and efficiently in a highly competitive global marketplace and manage new complexities in the information economy in ways that only computer- and communications-based technology can accomplish. Other drivers will be the need to establish better communication channels so that information can flow freely both within the organization and outside of it.

Perhaps the single most important infrastructure trend to emerge in the 1990s is internetworking: the proliferation of LANs and the emerging challenges pertaining to their interconnection. The deployment of PCs, their subsequent hookup in departmentally purchased PC LANs, and finally the attempt to internetwork these islands of LANs on a global basis was a naturally occurring phenomenon in the corporation, driven primarily by end-user requirements (see Figure 2.2). This stands in stark contrast to the big-systems approaches prevalent in the 1970s and 1980s whereby the installation of fixed, capital-intensive mainframe and minicomputer systems set fairly rigid limits on the kinds of capabilities that end users could

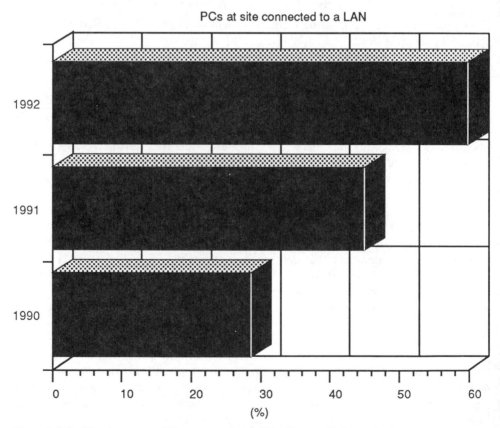

Figure 2.2 Profile of corporate PCs connected to LANs. (Source: IDC 1992.)

have. Unfortunately, under this approach, the real needs of the end user were often obscured by an overemphasis on the systems and technology, rather than a rigorous analysis of capability requirements and how best to fulfill them.

New LAN systems, therefore, created a different perspective on the development of computer and communications systems *beginning at the user level and moving outward*, as opposed to a structured approach imposed on corporate departments on a top-down basis. As this trend continues, more and more computer and communications intelligence is working its way to the periphery and end nodes of the network, meaning that the increased control migrating toward the end user will be further consolidated. As a result, as Figure 2.3 illustrates, new applications are being developed continuously. Furthermore, new router-based networks, along with next-generation "superhubs," are forming the basis for new distributed peer-to-peer networks that are, to a large extent, competing with and undermining traditional hierarchical SNA networks that had become the workhorses for most corporate computing. This new LAN internetworking model is also being developed and extended into the wide area, where new data-oriented transport technologies—such as frame relay and switched multimegabit data service (SMDS)—are being deployed by traditional VAN providers (such as BT Tymnet and Sprint) and the RBOCs, respectively. Moreover, these players, in addition to long-distance providers and alternate local carriers, are busy planning for the development of broad-

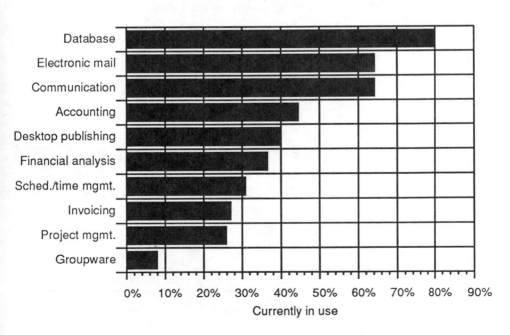

Figure 2.3 Commonly used LAN-based applications. (Source: IDC 1992.)

band ISDN (B-ISDN) capability throughout their networks. This technology will most likely be based on asynchronous transfer mode (ATM) technology, which is the CCITT-based standard for fast packet switching. (Figure 2.4 shows the relationship of these various transport technologies across an extended time scale.) Furthermore, even ATM-based B-ISDN, which can accommodate not only data traffic, but voice and video as well, may experience some competition in the future from commercialized Internet services, if such services begin to use switching technology (such as IBM's Planet switch) that is proprietary in nature or more closely aligned with TCP/IP networking than OSI-based approaches.

All of this new and emerging WAN technology is reinventing the market for data transport. The new LAN-WAN interconnection approaches, coupled with impacts of the client-server model, are serving in many instances to replace networking structures served by traditional large-scale hierarchical computer networks and systems. Unfortunately, in many cases, the result of this activity has been the deployment of two or more separate networks in corporate facilities, as discussed earlier. Needless to say, there are many reasons to try to reduce the duplication inherent in this situation. For one thing, running two parallel networks can be expensive. Furthermore, the two types of networks could hardly be more different. Networking under SNA usually involves the use of mainframe computers with front-end processors acting as the "traffic cop" for information flowing to various systems, cluster controllers to distribute communications further among small work-

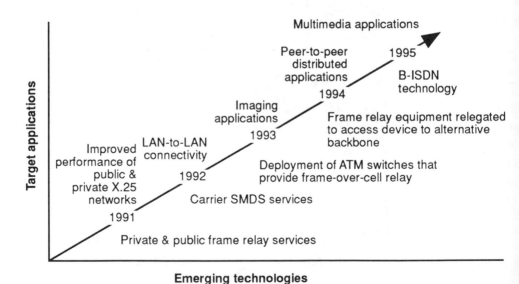

Figure 2.4 Time line: applications versus data transport capabilities. (Source: IDC 1992.)

ing groups, and 3270-type terminals sitting in front of the user. The new LAN-based paradigm, on the other hand, is increasingly populated with smaller, easily deployable, downsized (that is, appropriate to PC scale) equipment such as routers, bridges, communications servers, and smart hubs. Some of these products, as they fall in price and become even more commoditized, will attain almost "throwaway" status and hence will become widely used throughout the network by anyone who requires instant connectivity. This makes the "build up and tear down" process of these new LAN internetworks much more flexible and easy to accomplish within a relatively limited span of time, in turn making the new network paradigm far more flexible than the traditional network schemes developed under typical three-to-five-year MIS planning cycles. This has important implications for strategic advantage, because with the new types of corporate enterprises increasingly being developed—whether "virtual ventures" or Harvard Economist Robert Reich's "enterprise webs"—the luxury of these long-term planning cycles will no longer be available. To stay competitive, companies will have to learn to be "network-agile" and develop their strategic resources quickly and decisively to capture new market opportunities as they are identified [10].

There are differing views on how these two major network infrastructures for data will eventually become reconciled (see Figure 2.5). One likely scenario is that, as LAN internetworks—what Forrester Research has dubbed LINs—continue to grow in size and sophistication, they will eventually begin to accommodate the traffic that traditionally ran exclusively on SNA networks. (This, in fact, has already happened in many instances.) A key question in this scenario is whether SNA traffic will accommodate LAN traffic or whether router-based LINs will be able to adapt SNA traffic. Major router vendors how have SNA routing capabilities

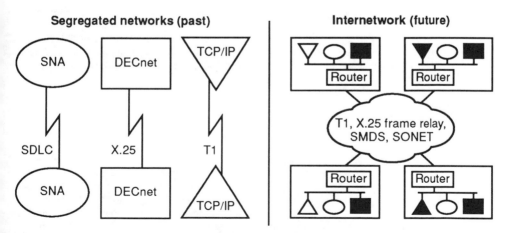

Figure 2.5 One scenario for collapsing parallel networks. (Source: IDC 1992.)

incorporated into their products, making the latter possibility more likely at this juncture [11].

All of this has significant implications for IS management. For one thing, they no longer have the luxury of depending on IBM's grand architectural schemes and the long-term planning and development cycles that are associated with it. If companies are expected to be more agile in the deployment of resources, then so is IS support for the various kinds of projects that might be required. In many cases, this will be a difficult transition because of the knowledge gap in many IS departments regarding the intricacies of these new LAN internetworks. The world of bridges, routers, IP networking, and the Internet is a culturally and technically different realm than the one that most IBM-oriented IS professionals are accustomed to. Furthermore, there needs to be a better understanding among all the players concerning the larger architectural issues pertaining to how these new networks will be integrated within more traditional systems. Accordingly, IS departments have retrenched by hiring more business-oriented talent, relying on the services of systems integrators to fill the technical gaps, reorganizing themselves toward the concept of a more decentralized service organization, and shortening their planning horizons from five years to two (or fewer) years. Although some observers see these trends as a downturn for the influence of future IS roles within the corporate organization, others see it as an opportunity for IS/telecom managers to become stronger and much more involved players in the development of networking-based strategic resources.

NOTES AND REFERENCES

[1] Peter Huber, *The Geodesic Network: 1987 Report on Competition in the Telephone Industry*, U.S. Department of Justice, Washington, D.C., 1987.

[2] *The Post-Divestiture U.S. Telecommunications Equipment Manufacturing Industry: The Benefits of Competition*, joint study conducted by the North American Telecommunications Association (NATA), the Independent Data Communications Manufacturers Association (IDCMA), and the Telecommunications Industry Association (TIA), 1991.

[3] *New Connections for the 1990s*, International Communications Association, Washington, D.C., 1990.

[4] Testimony of Mitchell Kapor before the the Massachusetts Department of Public Utilities.

[5] Private communication with Anthony Rutkowski, vice president and cofounder of the Internet Society.

[6] See John Markoff, "U.S. Said To Play Favorites in Promoting Nationwide Computer Network," *New York Times*, 18 December 1991. In the article, David Farber discusses the NREN and issues of Internet commercialization: " 'This is the first major communication business to be born under the deregulation era,' said David Farber, a computer scientist at the University of Pennsylvania and a pioneer in data networking. 'This hasn't happened since the growth of the telephone industry.' " The article continues: " 'I think it's a mess,' said Mitchell D. Kapor, the founder of Lotus Development Corp. and now head of the Electronic Frontier Foundation, a public-interest group focusing on public policy issues surrounding data networks. 'Nobody should have an unfair

advantage. This is important because we're talking about something that is in its infancy but that one day could be on the order of the personal computer industry.' "

[7] Scott Leibs, "Information: Plug in, Turn on, Tune up," *Information Week*, 12 November 1990.

[8] James Dart, "High Performance Computing for the 1990s," *Telecommunications*, January 1991.

[9] For more background, see Thomas S. Valovic, "Downsized, Do-It-Yourself Communications Technology Is Here to Stay," *Telecommunications*, December 1991, 6.

[10] "This is What the U.S. Must Do to Stay Competitive," *Business Week*, 16 December 1991.

[11] J. Hyland and M. Modahl, *Building New Networks*, Forrester Research, Cambridge, MA, 1991.

Chapter 3

How Telecommunications Is Transforming the Corporation

> Your enterprise is an information system which incidentally sells goods and services.
>
> —*Joseph F. Coates*

Telecommunications (and its subset, data networking) has become a radical force for change in the corporate environment. Not only has it enabled the organizational transformation of the modern corporation and its traditional management structures, it also has changed the fundamental nature of relationships in the corporate nexus of employees, managers, suppliers, customers, resellers, and support organizations. In fact, there is little in the contemporary business environment that the new realities of communications haven't in some way affected.

The massive deployment of computer and communications capability into the corporation happened at a time when major upheavals were already beginning to occur. To what extent, then, was this new technology responsible for the downsizing and restructuring of corporate organizations, as well as other entities? The answer to this question, of course, is a matter of some debate. We can, at least, surmise that computer networking had a significant impact on these changes and acted as an enabler or magnifier for changes that already had gathered momentum as the result of a confluence of other social, economic, and cultural factors. With somewhat less certainty, however, can we suggest direct and causal links between and among such phenomena as the restructuring of corporate management, the flattening of organizational hierarchies, experimentation with new approaches to developing and introducing products, and the reinventing of many aspects of the market distribution process. The degree to which communications is directly responsible for such changes is perhaps too academic a question to be explored here, so we will leave such considerations to others. Suffice it to say that computer networking, in ways not yet fully understood, represents a major force in the ongoing organizational transformation of the American corporation and, beyond this, that new information technologies are laying the groundwork for a new and innovative paradigm of workplace interaction.

As computers and communications became increasingly visible in the corporate setting, management reacted to the new technology in different ways. Some viewed this phenomenon as simply an extension or addition to the methods and procedures that had previously been accomplished manually; computers and computer networking simply allowed the same procedures to be done a little better, faster, or differently. Others saw the wider implication that this new technology had for the way their companies conducted business—namely, a largely unexploited and unexplored capability that could make a positive impact in many areas. In fact, in terms of its impact as a change agent, the computer revolution took many by surprise. Mired in their big-systems thinking, even the most influential purveyors of computer hardware were often unable to muster the requisite vision to see how the technology they were fostering would eventually give workers a powerful new set of desktop tools that would transform the marketplace and the workplace dramatically and inexorably.

This chapter will discuss some of the major business transformations that the strategic use of communications has now made possible. It will look at unique approaches to the development of new products and services, how new channels of distribution are being opened up for those products and services, how new communication media are creating and redefining the increasingly important area of aftermarket service and support in the emerging service economy, and how these media have contributed to blurring the line between products and services. It also will look at how these capabilities gave so-called "smokestack industries" (those that were visionary enough to seize the opportunity) a new lease on their business life—and the flip side, how other companies have failed to recognize the profound implications that information technology would have on them. Finally, we will examine how more forward-thinking companies are quickly moving to seize these opportunities by radically redefining their traditional business charter.

NEW APPROACHES TO MARKET DISTRIBUTION

One of the most interesting impacts of information networking technology on the corporation relates to the fundamental mechanism of market distribution. Once a product rolls off the assembly line, how does it find its way into the hands of the interested, paying customer? In the past, distribution channels, like management structures themselves, have been hierarchical: manufacturers dealt with wholesalers, wholesalers with distributors, distributors with retailers, and retailers with customers. Much of the market in any given retail industry was saturated with middlemen, who, after garnering their service fee, moved the goods to smaller outlets in various sizes. Because of this layering of additional (and perhaps unnecessary) channels between the manufacturer and the customer, distortion and poor communication could easily arise in the information flow necessary to direct the process of matching goods with consumer needs. This distortion often produced significant inefficiencies in terms of overstock, understock, and poor responsiveness to the needs of the customer. In this sense, before computer networking arrived on the scene, the retail marketplace was a supplier's, not a customer's, game.

The advent of information technology has fostered the dissolution of this cumbersome hierarchical structure. It has helped bring customers closer to the producers of goods and services and, in doing so, has made those producers more responsive to customer needs. It has also removed the seemingly superfluous pricing adjustments (that is, markups) that have customarily been built in to the cost of those goods and services to accommodate middlemen in the traditional distribution structure. These trends have had significant repercussions on wholesalers and distributors, as well as manufacturers and retailers situated at the extreme endpoints of the distribution chain. Furthermore, new and hitherto unavailable data about customer buying patterns is now flowing up that chain from consumer to producer, helping the latter to make far more effective marketing decisions and target potential buyers more accurately. The data itself has become a hotly contested kind of "information currency" among these once cozily cooperative but now highly competitive marketplace entities. Because the retail domain represents the closest point of contact to the customer, retailers have also come to appreciate that they are in the best position to gather and take possession of this highly prized information: exactly what products the customer is buying, and when. Just as computer networking has empowered users in the workplace, so is it giving consumers a voice in the marketplace, as companies scramble to gain possession of the inside track, characterized by the gathering of the right market intelligence at the right time. For every major player in the traditional distribution chain, this kind of information has become the new marketing currency. Companies that have this information—and use computer and communications technology to creatively process and package it—stand to benefit enormously.

How exactly have these changes taken place, and what role has new technology played in acting as catalyst and change agent? An excellent example in the retail sector are larger supermarket chains and conglomerates that gather information about customer purchases and transmit it back to a regional corporate computing facility, where it is processed, sorted, and collated before being transmitted via the company's internal network to yet another computer at the chain's main corporate facility. The now-familiar universal product codes and scanners make it possible for the store to gather data on each and every customer purchase. This accomplishes two things. First, it gives the retail store up-to-the-minute information on existing inventory and warns it when supplies are getting low. Second, it provides instantaneous feedback on the effectiveness of various marketing approaches, such as weekend sales, monthly specials, and so on.

The use of these kinds of techniques gave store managers for the first time a dynamic glimpse in near-real time of what was actually happening on the retail floor in terms of basic buying and selling transactions and trends. Even more importantly, it shifted the locus of power in the market distribution chain over to the retailer, now in possession of a sophisticated data base that reveals what customers want and when they want it. It allows retailers, for example, to negotiate far more effectively with individual suppliers, making them much less dependent on the major manufacturing conglomerates. It also creates a new constellation of suppliers around each store, eager to respond to the more finely tuned and specialized requirements that each store can now define, using the new technology.

A battle over the complex issues associated with the control of customer information is raging and will likely continue for some years to come. Because this information is so highly prized, it will be the locus of attention and competition between a number of entities in the retail marketplace. For example, when a credit-card transaction takes place, does the information about the purchase "belong" to the customer, the retailer, or the credit-card company facilitating the purchase? The answers to these questions, coupled with new concerns about privacy and protection of the consumer, will likely be a matter of public debate as they are played out over the next few years.

MOVING CLOSER TO THE CUSTOMER

Computers and communications have had major impacts on the traditional retail marketplace. For one thing, they have allowed both manufacturers and retailers to more closely identify marketing targets (that is, to whom they are selling their products), removing much of the guesswork from the development of their product offerings. The derived benefit is that, knowing the marketplace more thoroughly, manufacturers expend less effort developing products and services that will be "trial ballooned." The advantage for the consumer is that these providers are becoming far more responsive and competitive with respect to their needs. Furthermore, increased competition means lower costs and better service.

However, this is just the beginning. The next phase will move the manufacturer even closer to the customer and, in some areas of retailing, may threaten to pigeonhole even the retailer itself into the increasingly undesirable category of "middleman." This phase relates to the use of home-based electronic information services (see Chapter 4) that will provide even more direct connections between manufacturer and consumer. Prefiguring this trend has been the meteoric rise of the mail-order catalog business, whereby companies can compete directly for a consumer's attention, while the consumer need never set foot in a store. Interestingly, such approaches give the manufacturer (rather than the retailer) direct access to information regarding customer buying patterns.

Companies are already gaining more access to customers via the use of electronic information services such as Prodigy and CompuServe. Prodigy, for example, allows users to electronically access a wide variety of products and services, including the SABRE system for airline reservations, Sears catalog shopping, and an array of retail outlets and systems. In the realm of broadcasting, which will eventually converge with electronic information services, businesses such as the Home Shopping Club now compete directly with large merchandisers such as Sears and indirectly with a wide spectrum of individual retail outlets and specialty stores. These trends are expected to accelerate now that the RBOCS have been given the regulatory go-ahead to deploy advanced information services in their respective serving areas throughout the United States.

The battle for control of newly emerging and more streamlined distribution channels means that various entities involved in traditional distribution are often forced to become competitors, whereas previously they were simply comembers in the distribution chain. Although manufacturers theoretically have the upper hand in providing a base of products for distribution, various entities in the chain, such as distributors themselves, are now free to leverage their own position through the strategic use of telecommunications. A classic and oft-cited example of this phenomenon is American Hospital Supply (now part of Baxter Health Care Corporation). What AHS did, using networking technology, might be described as a "Trojan horse" strategy. The same approach was used successfully in a number of other marketing coups, including American Airlines' deployment of the SABRE system for airline reservations. The idea is simple in both design and execution: place direct ordering terminals on your customers' premises, allowing them direct access to vital information about availability, inventory, shipment dates, custom options, and so on. Thus the sharing, not withholding, of information becomes the principal means of developing sustained competitive advantage. The approach taken in both instances involves giving the customer more control of the company's own data resources. In return, the supplier benefits from a customer that is "locked in" to the advantages of using a value-added service, over and above the simple mechanics of product delivery. The service in this case becomes the ability to order supplies in a significantly more efficient manner, allowing AHS customers to keep inventories much leaner, thereby saving time, labor, and storage costs. More importantly, however, what AHS sold to their customers was a management system that became part and parcel of the customers' internal operation. Thus, a new form of cooperative relationship was formed, which, as will be discussed later, represents a major transforming agent in the new global marketplace [1].

Thus, companies that can provide these information conduits to future consumers, thereby bypassing traditional market chains, will gain a decided advantage in the virtual electronic marketplace of the future. Accordingly, it's not too difficult to project that the battle over who ends up with control of these conduits will become one of the most critical and fiercely contested issues in the strategic use of telecommunications throughout the 1990s. One such battle is already taking place among the RBOCs, CATV providers, and the publishing industry over who will gain access to potentially highly lucrative residential markets via newly emerging entertainment and interactive information services [2].

The use of electronic communications also has the unique ability to affect the way corporations utilize physical resources and space. Through the use of more accurate and real-time data on customer requirements and shipments, companies are now able to reduce the need for resources traditionally used to buffer the lack of knowledge about such factors. For example, before the advent of real-time in-process information on customer purchases, seasonal buying patterns, and sales as a function of particular economic conditions, manufacturers were required, via warehousing, to overstock their product in anticipation of possible aberrations in the normal selling cycle. With computerized information available to describe these patterns, companies can operate with far less physical storage in adherence with the now widely used principles of just-in-time (JIT) manufacturing. The net result is a more customized approach to product design and development.

The principal benefit of this reduction in company operations is obvious: a general streamlining of the inventory process. The elimination of warehousing requirements means that less physical facility space, personnel, and equipment are required to maintain the inflow and outflow of product into company facilities. Some economists and experts in the field of social organization, in fact, see this trend very much related to a general trend toward corporate downsizing. In this new scenario, specific company functions once subsumed under one roof will increasingly be handled by a constellation of support organizations clustered around that company. What will enable this decentralized approach is a comprehensive ability to connect these functions via computer networks. *Thus, one of the most important impacts that electronic communications is currently having on corporations is to bestow the capability of both time and distance insensitivity.* This means that the corporation can draw upon resources wherever and whenever it needs them, as opposed to being heavily dependent on the somewhat unpredictable cycles of traditional resource environments.

Such approaches to the marketplace are keyed to several basic factors. Having the processes of business transaction tied to the ordinary constraints of time and space in the past implied several built-in limitations. For example, it meant that the market for many companies was geographically restricted. However, new approaches made possible by computers and communications now mean that markets can be, in theory, almost indefinitely expanded. Up until fairly recently, companies were limited to selling products to one segment of the country during a fixed time period via a retail delivery system. Under the new networking realities, markets can now be expanded to become nationwide, or even global, in scope and scale. These systems create time as well as distance insensitivity, engendering global, 24-hour, multidimensional marketplaces. Many direct-mail companies have become extraordinarily successful as a result of recognizing these new realities. A classic example of a business that now operates 24 hours a day, 365 days a year, is the well-known mail-order house L. L. Bean (a company that, incidentally, accounts for more than half of all the telecommunications traffic coming into the state of Maine, where its corporate facility is based!). In addition, these two factors are related. Companies that operate 24 hours a day, for example, can take advantage of greater geographical flexibility, because their business can now span time zones.

Thus, the newly emerging market reality is predicated on a simple fact: vendors can now actively *search* for a customer interested in buying their product, as opposed to the now, by contrast, seemingly lackadaisical approach of hanging out a shingle and waiting for customers to drift into a retail outlet or other product delivery system. Electronic marketing makes this all possible by creating virtual marketplaces that bring buyers and sellers together in new and unique ways, a trend that has only just begun to manifest itself in the highly charged, globally competitive market of the 1990s.

REINVENTING CORPORATE MARKETING

What exactly is marketing? In the classic definition, marketing involves "directing the flow of goods and services from producers to consumers or users" [3]. In a variety of business areas, computer networking has transformed the marketplace in unexpected ways. These changes and their aftershocks are still taking place and are far from reaching a tranquil or static end point. They have fundamentally altered market relationships between customer and supplier, distributor and manufacturer, wholesaler and agent. But probably no other area is as much a major beneficiary of such change as the marketing function itself.

As has been described, a major transforming agent in marketing has been the ability of computer networking to help companies find customers instead of waiting for the customer to find them—an active instead of passive approach. In addition to enabling this aspect of the marketing process, sophisticated computer and communications methodologies also allow the vendor to *narrow the probability of the sales transaction* by zeroing in on customers most likely to need specific items that they have to offer. Thus, sophisticated analyses of buying patterns allow companies to presell as well as sell products and services. Preselling in this sense simply means that, in a nexus of complex social and business events, an established and often recurring need is identified and steps are taken to ensure that the company has their product available in the right place at the right time.

The critical dependency here is a simple one: establishing a predictable data base of buying patterns. One way to do this, of course, is to extrapolate this data from a historical data base. A soft drink supplier, for example, could identify a series of dates when a local sports facility will need concession stand stock. But more important than simply using a static data base that lists special events would be to plug in information pertaining to specific adjustments—for example, playoff schedules leading to an extended sports year. This is current and near-real time information that no amount of computer smarts can readily extrapolate. Instead, what's needed is knowledgeable manipulation of a data base by analysts who can project marketing patterns and build this information into the system on a case-by-case basis.

In the retail arena, these techniques also apply. Domino's, for example, maintains a computerized data base of their local customers. When a customer first calls in to have a pizza delivered, the local franchise will gather pertinent information to establish a customer profile, including the customer's address, method of payment, and so on. When subsequent calls are made, the necessity of repeating this information is eliminated. The net effect is that the entire transaction is made easier and more pleasant for the customer, which helps ensure repeat sales, in part by creating a sense of continuity and familiarity.

Wrongly applied, however, the new competitiveness that computers and communications bring to the marketplace may also be perceived as a form of over-aggressiveness on the part of vendors and suppliers. At the corporate or business-to-business level, this may not be as problematic as what can transpire at the level of the individual consumer. At stake are a host of privacy issues already coming to the forefront as more sophisticated marketing techniques intrude on the rights of consumers to newly emerging concepts of informational privacy. Whereas it might be one thing for a supermarket supplier to anticipate buying patterns for a chain of supermarkets, it's another entirely for a company to accumulate data on the purchasing patterns of an individual without his or her informed consent or awareness. Once this kind of information has been garnered, it is difficult for companies to resist taking the next step, which often may involve what some might consider to be intrusive marketing measures to capture that consumer's attention.

As the virtual electronic marketplace matures, one of the major sources of information about consumer purchases will be gleaned from the use of smart cards and credit cards. These transaction instruments are increasingly being used to identify specific purchases made by the card holders. If this information is used only by the party that gathers it, then its use seems legitimate, and problems are unlikely to arise. But when this information is sold or otherwise distributed to third parties, many troubling public-policy issues can surface, and vendors are well advised to think them through carefully before they inadvertently make enemies of the very customers they are trying to court.

THE "USER AS VENDOR" PHENOMENON

In addition to changing traditional marketing distribution patterns, new networking technologies are also effecting radical change in a number of other ways. For example, traditional smokestack industries are actually stepping away from their mainstream business operations and electing to become direct players in emerging markets for telecommunications or information services.

In considering this phenomenon, it's important to remember that every major *vendor* in the Fortune 1000 and beyond is also a highly experienced *user* of information services. Among IBM's very biggest customers, for example, are the Bell operating companies. In turn, one of AT&T's biggest customers is IBM. Thus, many companies are users as well as vendors in the same market sphere. This may not seem unusual; however, an interesting market dynamic develops when companies begin to leverage the user dimension of their operations by extending those capabilities to the vendor side. Digital Equipment Corporation is one of the prime examples of this unique kind of cross-fertilization. DEC is a company that has over the years learned to develop extraordinarily good internal information transfer and synergy between their in-house telecommunications functions and the core business operations responsible for their product development process. For example, the company often uses its internal network as a test-bed prior to the launch of new customer products. Another example is when DEC made front-page news in the trade press by developing an outsourcing arrangement with Eastman Kodak. This in and of itself was not particularly newsworthy, because DEC has routinely provided systems integration to a wide range of companies. However, what made this a watershed event was the fact that DEC provided network management services to Kodak not for data but for voice communications. Even more significantly, the reason they were able to do this was that they had developed such sophisticated network management approaches for their own operations. Turning around and marketing that expertise to outside companies then became simply another logical step in the process.

In some cases, companies have made strategic moves into information or communications services as a result of their in-house capabilities spilling over into other areas of their business strategy. In other cases, companies were presented with new and unusual opportunities as a result of changes in the marketplace. In any event, one of the most fascinating aspects of the evolution of telecommunications capability in the postdivestiture era continues to be its ability to transform and redefine the very nature of the Fortune 1000's traditional business charter. Consider, for example, the following:

- In the late 1980s, Sears decided to market its own extensive, nationally deployed, private network to outside companies. The entity created to do this and resell the excess network capacity was Sears Communications Corporation. The subsidiary could frequently be seen displaying its wares at major industry trade shows.

- In 1991, Mobil Oil Corporation began promoting itself as a provider of audiotext services. These were primarily 900-number-based services that callers could dial into for information pertaining to news, sports, weather, or other user-interest categories.
- American Express announced the formation of a separate business unit for the purpose of launching itself into the videoconferencing business in 1991. In order to do this, the company developed a series of alliances with other telecommunications service providers.

While corporations throughout a broad spectrum of vertical industries unrelated to telecommunications were deciding to get into this dynamic new market, there was another side of the coin: companies already in the computer or communications business migrated their operational orientation steadily toward vertical market applications. (In a sense, this was simply the mirror image of what many companies outside of the communications field were trying to do. However, the end point was often the same.) This phenomenon has parallels in the way in which software systems eventually came to be marketed in the computer industry. For example, in the 1980s, there were a considerable number of new companies whose primary charter was to develop and market software for specific vertical industries. An example is Meditech, a software house specializing in the health and medical markets. As the nature of business becomes increasingly computer- and information-oriented, the actual practice of the business can eventually become indistinguishable from the computer operations that support it.

Similarly, the move toward vertical market emphasis is starting to become evident in the communications arena. In some cases, a company develops telecommunications software or systems in-house and, finding that the product has a broader appeal, markets it externally. Traveler's Insurance is an example of one company that marketed its own communications software through a separate subsidiary called Travtech. In other cases, orientation toward vertical markets has occurred as the direct result of higher levels of competition in certain market segments. The PBX industry, for example, ran into difficult straits in the late 1980s, as shrinking margins forced vendors to compete more creatively in the marketplace. One way of doing this was to develop more feature-rich switches. Another was to custom-tailor them via software enhancements that addressed the specific business applications of customers. NEC, for example, developed applications packages for its PBXs, which target the hotel industry, largely as a means of product differentiation. The software offered features with designations such as Suite Room service, VIP automatic wake-up service, and Alert service.

One of the more interesting synergies that has developed is that between an old smokestack industry, the railroads, and a new information age business, fiber-optic communications. Interestingly, this link goes back to the earliest days of communications history when the first telephone and telegraph lines were put into national use. With the advent of fiber optics, the railroad companies again began using their rights-of-way to create what was to become a major portion of the nation's best and newest fiber-optic networks. Amtrak, for example, is now in the transmission business and frequently exhibits at communications trade shows. Sprint has its business roots in the railroad business. Following Sprint's example, many other companies seized the opportunity to use their existing rights of way as a means of entering the communications market, placing themselves in competition with (at the time) the older microwave-based long-distance companies such as AT&T and MCI. Likewise, the Williams Company, a company involved in the oil business, became a major alternate long-haul purveyor of fiber-optic capacity using the rights of way for their pipelines. WilTel, their telecom subsidiary, started out with 900 route miles of fiber-optic transmission capacity in 1985 and now has over 11,000 route miles in a network that covers major portions of the United States. This newly minted independent transmission capacity has now formed the basis of new alternate long-distance networks that have helped drive prices down and created new levels of competition in an industry once quietly controlled by one company, AT&T and its Long Lines division.

Thus, what we seem to be witnessing is such a strong move toward the use of information as a competitive tool that many companies are no longer perceiving themselves as categorized within a specific vertical industry; rather, they are beginning to view themselves as information providers *specializing* in certain vertical market areas. This is arguably one of the most profound business-oriented paradigm shifts of the century.

This same blurring of business charters is evident in the battles being waged between the publishing industry, CATV operators, the RBOCs, and independent information providers, where each of these businesses perceives the others to be on a collision course with itself. When publishing becomes electronic, the question becomes: who will be best positioned to provide those services, and who will the regulatory climate favor? Intensified by the process of divestiture, these industries have been locked into what seems to be an endless struggle over dominance of the emerging electronic marketplace. The result has been a decade-long stall in the deployment of those kinds of services, only recently alleviated since the MFJ-mandated information services restrictions have been lifted from the RBOCs.

At the center of this battle is electronic information services, but the underlying significance is the simple fact that electronic text and media have the potential to surpass or at least equal the market dominance of traditional publishing media. The amount of revenue to be made in providing such services is staggering. As a result, the RBOCs, the publishing industry, and the cable providers have for many years engaged in massive lobbying to prevent each other from establishing a preemptive foothold. Each of these providers, however, approach this brave new world of electronic text from different vantage points. The RBOCs have traditionally been pure transmission providers, prohibited from content-based service by virtue of the their stranglehold on local exchange services. The publishers, of course, are completely content-driven. The cable companies, on the other hand, fall somewhere in between. But the point is simply that, as these new businesses converge, they will increasingly have common elements and begin to look more and more alike in their business charters. (In all likelihood, there is room for all of these industries as major players, hence the irony of the four groups racing to get through a revolving door that will admit only one!)

NETWORKING AND ORGANIZATIONAL CHANGE

Computer networking has become a major critical dependency in the restructuring and downsizing of corporations. This has been extensively documented in a number of sources, including Alvin Toffler's *Powershift*, which discusses several major corporate paradigm shifts either directly or indirectly related to newly emerging communications technologies. Toffler, in fact, appears convinced that the computer and communications revolution has indeed been a causative factor. Furthermore, he cites the death of the smokestack economy as simply a recognition that large-scale bureaucratic organizations have reached their own predetermined limits to growth. In other words, real economically viable and organizationally robust growth at some point became no longer possible without changing the very nature and structure of the corporate organization.

> . . . There is mounting evidence that giant firms, backbone of the smokestack economy, are too slow and maladaptive for today's high-speed business world. Not only has small business provided most of the 20 million jobs added in the U.S. economy since 1977, it has provided most of the innovation. Worse yet, the giants are increasingly lackluster as far as profits go, according to a *Business Week* study of the thousand largest firms. [4]

In another study of rapidly changing corporate cultures, *The Work of Nations*, Harvard economist Robert Reich cites specific numbers supporting Toffler's observation:

> . . . By most official measures, America's 500 largest industrial companies failed to create a single net new job between 1975 and 1990, their share of the civilian labor force dropping from 17 percent to less than 10 percent. Meanwhile, after decades of decline, the number of people

describing themselves as self-employed began to rise. And there has been an explosion in the number of new businesses (in 1950, 93,000 corporations were created in the United States; by the late 1980s, America was adding about 1.3 million new enterprises to the economy each year). Most of the new jobs in the economy appear to come from small businesses, as does most of the growth in research spending. [5]

As companies continue to downsize, the new organizational traits that will become *de rigueur* for highly competitive, globally adaptive multinationals in the 1990s will be, among others, flexibility and innate capability to increase speed to market with custom-designed and market-responsive products and services. This will require redesigning patterns of information flow and access throughout the corporation. As most observers of organizational behavior know, this cannot be accomplished without parallel changes in the structure of the organization. *This fact of corporate life is based on the principle that information flow in and of itself is the prime organizing principle in any organization.* Thus, installing a LAN in a company division can become a force for change that shifts the patterns of information flow and, hence, the power associated with those patterns.

Toffler also discusses the formation of new, smaller groups of companies dedicated to task-specific projects, a common variant of which are so-called "skunkworks." This is a business unit in which entrepreneurial teaming and problem solving in working toward very specific goals is encouraged and facilitated. Toffler goes on to define yet another *ad hoc* corporate organization that takes the concept of optimizing group initiative one step further: the "self-start team." Fundamentally antihierarchical in nature, these are clusters of individuals within the corporation that elect to organize themselves to take on new tasks or project proposals. *Such ambitious attempts at new forms of corporate organization, however, are arguably impossible to achieve without the use of the capabilities provided by the new realities of computer networking.*

Actually, while these ideas have lately found favor in the general business press and seminar circuit, there are organizational precedents for such approaches. One, for example, can be seen in the organizational structure of the modern management consulting firm. As a member of the client staff at Booz Allen and Hamilton, I was intrigued to learn how such firms really operated: for the most part, on the principle of chaos dynamics! Management consulting firms tend to be, organizationally, nothing more than a loose aggregation of highly motivated and entrepreneurially oriented consultants—hired guns, if you will—who, on an ad hoc basis, market their capabilities either individually or in small teams to other companies. They do so under the nominal aegis of a Booz Allen, an Arthur D. Little, or a McKinsey. But the activities they engage in are largely self-organizing and based on individual initiative, rather than in response to prodding from higher management levels. Furthermore, the teams tend to be highly cross-disciplinary in nature, at least among the more technical firms. The entrepreneurial pressure to remain "billable" is considerable, such that if a consultant's year-long project is drawing to a close, it behooves him or her to begin looking for a new project. A consultant's job, therefore, is not a job at all, in the conventional sense, but consists primarily of a series of projects or tasks. In this sense, management consulting and how it has been traditionally practiced might provide some clues as to how the knowledge workers of the future—what Robert Reich calls "symbolic analysts"—will structure their professional activities.

Extend the consulting model on a wider basis throughout newly emerging corporate environments and you get an interesting result. Via computer networking, this *ad hoc* approach might span many different types of organizations, leading to a concept of "employee as consultant." It's not difficult to imagine project teams being established on a cross-organizational basis—perhaps involving staff from both academia and industry—or even arising from the ranks of specific professional organizations whose members are linked electronically. IEEE members, for example, might use a bulletin board system that catalogs their specific technical areas of expertise to develop specialized projects and ad hoc teaming for fixed short-term cycles. Thus, communications might become the enabling principle for allowing motivated individuals to initiate a variety of projects, rather than waiting for an institutional entity to seek them out (which, of course, may not happen). Even more importantly, the projects themselves, given the eventual availability of advanced information services at the residential level, might in many cases actually be conducted electronically or "virtually." (This is one twist on the concept of "virtual ventures," an idea that will be discussed more thoroughly in later chapters.)

Stretching the imagination a little further, it's possible that such virtual workplace conventions will develop along the lines of competitive bids. For example, if a company needed a specific technical problem solved, they might conceivably bid it out to thousands of engineers throughout the IEEE by posting the parameters of the task on a bulletin board. Then, through the wonders of computer networking, the technical challenge would fall into the laptop of one of several people in a specific region who is best qualified to solve the problem. *Thus, just as computer networking is allowing vendors to actively search for customers, it will also allow employees to search for employers (and vice versa) in the realm of short-term, specific projects.* Needless to say, the ability to find precisely the right person to perform a specific assignment has very significant implications for productivity, especially if distance insensitivity is factored into the equation.

THE PHENOMENON OF "VERTICAL DISINTEGRATION"

As corporations seek to restructure themselves in order to become more competitive, they are shedding many of what were once considered core functions by parceling them to outside agents. As this happens, it is the strategic use of communications that keeps this centrifugal force from tearing asunder the organizational stability necessary to keep the company functioning as an integrated unit.

Behind this trend is a process that might be described as "vertical disintegration." This phenomenon began to manifest itself in the early 1990s, propelled by a variety of market and economic forces but also fostered and facilitated by computer networking. In its simplest sense, the term relates to the fact that many companies are revising their traditional view of the ideal organization as a hierarchical, multilevel "layer cake" with various strata representing individual corporate functions or departments. Under the old corporate model, the true vertically integrated company was, in essence, a self-contained entity, with mutually supporting but distinct departments whose purpose was to support mainstream business operations. Many Fortune 1000 companies, for example, had an in-house market-research staff or advertising department. This is not to say that individual firms haven't chosen, at one time or another, to outsource these kinds of services in one fashion or another. Some firms might have elected, for example, to disband their training and education department in favor of having an outside firm provide the service. Thus, the process of outsourcing various corporate functions is not new. What is new is the fact that, in the late 1980s and early 1990s, many companies that traditionally maintained high levels of vertical integration began to outsource or seriously consider outsourcing these functions to outside entities. Thus a trend was born; and at the heart of this trend was networking technology, the glue that held these arrangements together and made them possible.

To what extent these trends might be viewed as the failure of one of the trendier management theories of the 1980s, so-called "intrapreneurship" might be a topic of lively debate. Certainly, the move toward creating separate specialty organizations during that period produced some interesting results. Skunkworks projects such as IBM's PC development in Boca Raton, Florida, are now well-known examples of the kinds of successes that are possible. But even companies that allowed themselves to become splintered into multiple independent profit centers often found that this approach was effective only up to a point. Perhaps— or so we can speculate—this approach simply reached the upper limit of its capacity to effect systemic change within the large corporation, at which point, even deeper changes were called for.

At least one emerging model for vertical disintegration is that of a large company at the center of a complex constellation of outside suppliers, who, in turn, themselves work for many other companies, both large and small. As more and more companies elect to outsource various functions, this spin-off effect should continue to create a strong stimulus to the emerging entrepreneurial economy. This should carry even further the trend toward individual entrepreneurship that began in the 1980s when, according to the now familiar statistic, more small businesses were created than at any other time in recent history. Throughout this process, the often-underutilized energies of in-house functions will be handed over to smaller companies that can deal with them far more responsively and efficiently on an ad hoc, contractual basis. The antibureaucratic effect will both serve the large company that outsources its activities and create scores of new entrepreneurial ventures, bringing more dynamic levels of competition into the marketplace and continuing to radically alter traditional business relationships.

To the extent that computer networking now makes it so, geographical proximity is becoming less and less a factor in the selection criteria of service and support firms constellated around larger companies. Furthermore, as this process continues to evolve, the emerging model for even "regular" corporate employment in the future may begin to increasingly take on the characteristics described above. Along with companies engaged in a multitude of contractual, outsourced agreements with smaller, entrepreneurial firms, there may also be a rise in the use of agreements with individual employees. These individually negotiated agreements may then constitute a complex mosaic of corporate workforces based on retaining employee-consultants, each with his or her own highly specific working arrangements, keyed to mainstream business operations.

In sum, the employee-employer relationship will likely become a far more individually negotiated phenomenon, allowing for individual personal and professional employee differences. However, with it will come a new set of problems and challenges for management: namely, how should this array of highly individuated and nonuniform relationships be effectively managed? Or to put it somewhat less optimistically: can it be? And, assuming that it indeed can and must be managed, what are the major challenges in doing so? As Charles Savage, in *Fifth-Generation Management*, describes this process, senior managers in the new environment need to understand that

> . . . their biggest challenge is to manage complexity rather than just cost and time. The swirling multiple interrelationships, both external and internal, are often more chaotic than orderly. Only if the creative abilities of the people—employees, professionals and managers—in the firm are unleashed can they expect to respond effectively to the multiple challenges of the market. Finely tuned bureaucracies with carefully defined policies, procedures, and job descriptions are no match for the next decade. They are too confining and rigid and are always out of alignment with the market. They cannot maintain a creative dialogue with their suppliers and customers. [6]

Note that the word "employees" might easily be added to the last sentence as well. In this trend toward vertical disintegration, there is an inherent paradox: although traditional organizations *seemed* more coherent in the sense of being tightly controlled, managed, and physically integrated, they were, in actuality, much less so than the evolving kinds of fifth-generation management structures expected to become widespread in the years ahead. Thus, as organizations become looser and more "interstitial" in their widening circles of association, new and daunting management challenges will continue to emerge. *Moreover, these challenges will be surmountable only via the use of new enterprisewide network computing structures.* In Savage's view, these challenges principally include addressing the old industrial-era problem of fragmentation; establishing new systems for accountability in newer, more dynamic organizations; supporting the focusing and coordination of multiple cross-functional task groups; and addressing the need for ongoing learning in the dynamically changing corporation.

Although not all of these challenges are addressable via new computing and networking arrangements, many are. For example, new groupware systems (discussed in Chapter 4) can target all of these areas to some degree. The same can be said for various computer conferencing systems that mimic many of the functions of groupware or, in fact, have prefigured these functions but have generally received less industry attention for a variety of reasons that will be explored further.

Savage then goes on to discuss what he considers to be one of the central problems in current corporate management structures and offers an unusual and insightful thesis that aligns five generations of computing with five generations of management, as shown in Figure 3.1. Based on this extrapolation, analysis of current organizational structures leads to the following conclusion: *corporate management structures have failed to keep pace with the generations of computer technology that are driving organizations into new patterns of dynamic change.* Furthermore, as shown in the diagram, third-, fourth-, and fifth-generation computer technologies are typically "being grafted onto second-generation management" [6].

With these trends as background, an important question emerges: exactly how should newly emerging computer and communications technologies be engaged to tackle the management challenges outlined, such that they can become integrated into existing systems as a vital dimension in the new corporate paradigm? As mentioned, groupware and computer conferencing systems are just one example of systems that can be deployed strategically within this paradigm. These are communications-based computer software systems that allow complex multiparty tasks to be managed effectively. Executive informations systems (EIS) and decision support systems (DSS) are other examples. As these technologies evolve and become more communications-oriented, they are likely to begin to converge to a certain degree. Such tools will become absolutely indispensable for a new breed of managers who will be called upon to manage relatively simple tasks with complex sets of professional resources or will be involved in parceling out complex projects among complex resources. The special characteristics of communications-based groupware—time and distance insensitivity, among other features—will allow this kind of "remote" management to take place. Without them, such changes in the corporate organizational paradigm might not otherwise be possible.

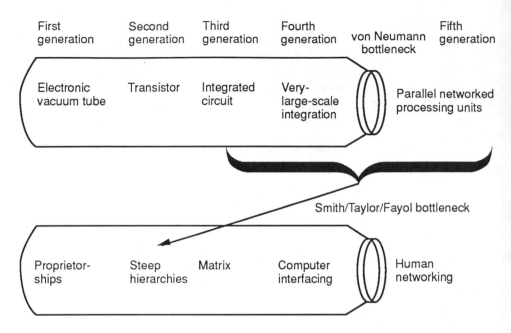

First generation — Second generation — Third generation — Fourth generation — von Neumann bottleneck — Fifth generation

Electronic vacuum tube — Transistor — Integrated circuit — Very-large-scale integration — Parallel networked processing units

Smith/Taylor/Fayol bottleneck

Proprietor-ships — Steep hierarchies — Matrix — Computer interfacing — Human networking

Figure 3.1 "Grafting" third-, fourth-, and fifth-generation computer technology onto second-generation management. Printed with permission from *5th Generation Management*, by Charles M. Savage, © 1990 Digital Press/Digital Equipment Corporation, One Burlington Woods Drive, Burlington, MA 01830.

NOTES AND REFERENCES

[1] For a good account of American Hospital Supply's approach, see Peter Keen, *Competing in Time*, Ballinger, Cambridge, MA, 1988, pp. 55–61.
[2] For more information on this subject, see Thomas S. Valovic, "Conflict or Cooperation: NREN and U.S. Telecom Policy," *Whole Earth Review*, Spring, 1991.
[3] Encyclopedia Britannica, s.v. "marketing."
[4] Alvin Toffler, *Powershift: Knowledge, Wealth, and Violence at the Edge of the 21st Century*, Bantam Books, New York, 1990, p. 181.
[5] Robert B. Reich, *The Work of Nations*, Knopf, New York, 1991, p. 95.
[6] Charles M. Savage, *Fifth-Generation Management*, Digital Press, Bedford, MA, 1990.

Chapter 4
Optimizing Internal Corporate Communications

Networking is contagious.

—Peter Huber

The term strategic use of telecommunications seems most often to conjure up a variety of customer-oriented applications at the vanguard of new information strategies targeted toward the emerging global marketplace. Considered in this context, strategic communications can be used to transform and alter traditional means of transacting business so as to reach new markets in new ways, reach old markets in new ways, create new markets, or simply improve the overall process of customer communications and feedback involved in product or service delivery. An equally important but less commonly appreciated aspect of using communications strategically relates to its use and deployment in enhancing a company's internal operations, particularly as applied to the major elements delineated in Porter's value chain (see Chapter 1). This is especially relevant given the widely accepted axiom that 80% of a company's communications are internal rather than external. The payback objectives associated with using new computer networking capabilities in this context include increasing efficiency, morale, innovation, speed-to-market, group awareness, corporate awareness, and a host of other factors. All of these have very real impacts on the ability of a company to stay competitive in the new information economy.

Prior to the so-called "information age," communications in the corporate world took place in one of four ways: personal meetings, group meetings, telephone communication, and written correspondence. Prior to the advent of paper copiers, these kinds of communications possessed what now, with the benefit of hindsight, appear to be an interesting and unusual quality: when such traditional forms of communications occurred, they tended to be unique and irreproducible. Thus, a

---

telephone conversation, memo, or meeting had the quality of existing in a unique locus of space and time and, as a result, the information they contained was generally not widely distributed. Such information remained tightly controlled within the corporate jurisdiction responsible for maintaining it. It was, in a sense, private information. In line with this scenario, the traditional middle manager fulfilled the role of information filter, custodian, and, when appropriate, distributor, acting on behalf of his or her department or division within the company.

The advent of office automation technologies, with their propensity for wider information dissemination, placed a radical new spin on this scenario. Photocopy machines allowed memos and correspondence to be copied and distributed to a much larger audience. Facsimile machines allowed memos and letters to be distributed more readily outside of the corporation. However, the enabling technology that had the greatest impact on how business was conducted both internally and externally was the personal computer. PCs initially offered fairly mundane capabilities such as word processing, but they also offered the ability to store and manipulate large amounts of information—another step toward destroying information's quality of uniqueness.

J. THE PC AS A LEADING-EDGE TOOL

As PCs became more widely deployed, they began to change the patterns of traditional intracorporate communications. As information flowed more readily across the increasingly permeable membranes of sometimes arbitrarily divided business functions, it forced many glitches and inefficiencies into the scrutiny of corporate management (including many of the practices of management itself). Stewart Brand's axiom that "information wants to be free" thus proved accurate as leading-edge users in the corporation—realizing the benefits in doing so—began sharing information, and more enlightened managers worked to encourage the process via new procedures and technology. This process of communication created new organizational dynamics, the synergies of which have yet to be fully documented.

Each of these technologies had its own particular contribution to make to the creation, storage, processing, and distribution of information. In many cases, these functions were closely related. For example, facsimile machines, photocopiers, and laser printers all, in essence, perform a similar function: taking an information input and converting it into hard-copy text. All of them had significant impacts on internal and external communications. However, as processing power becomes even cheaper and the technical capability to encapsulate more and more functionality into less and less microchip space is enhanced, these stand-alone devices will began to disappear. As such, they are transitional technologies on the way to far more integrated capabilities. In this respect, all of the technologies mentioned above have prefigured the next wave of office communications: the consolidation of all of these functions at the desktop using a single device and a single communications line.

Rather than dwelling on the effects of these transitional technologies on internal communications, the remainder of this chapter will focus on newly emerging PC-LAN-based office communications systems, which will likely have the greatest overall operational impact in the new electronic landscape. These include

- Electronic mail;
- Computer conferencing;
- Commercial use of the Internet;
- Electronic information services;
- Groupware;
- Telecommuting.

ELECTRONIC MAIL APPLICATIONS

From a strictly technical standpoint, electronic mail has been around for a long time. However, its acceptance and use in the corporate environment has emerged only gradually (see Figure 4.1). In some environments, e-mail existed as simply a feature of a larger, more diverse office communications system such as DEC's All-in-One system or IBM's PROFS. These desktop utility systems were used for a variety of purposes, and e-mail represented one capability among many (and was often considered a novelty by users). The real key to the deployment of e-mail in the office has a lot more to do with the somewhat trickier phenomenon of cultural acceptance than the specific availability or application of the technology. A number of factors are likely to have been involved in those cases in which the technology was available but was simply not utilized, including lack of critical mass among users, lack of full support from management, technical difficulties that diminished ease-of-use, and inadequate training or systems introduction.

A good example of the importance of ease-of-use and compatibility is evident in systems in which e-mail cannot be accessed on an in-session basis but requires the user to quit a given session and, in a somewhat cumbersome fashion, plow through multiple screens in order to arrive at one that affords e-mail access. (The recent increased popularity of PC-based windowing via such products as Microsoft Windows has helped to alleviate this kind of disincentive.) Nevertheless, the problems of poor ease-of-use inherent in early releases of computer software were often enough to make e-mail sufficiently user-hostile that, in many instances, critical mass was never able to develop.

Organizational factors and constraints also have major impacts on the use of e-mail. Generally, the more hierarchical the organization, the less likely e-mail will become an acceptable and ubiquitous form of communications. E-mail and

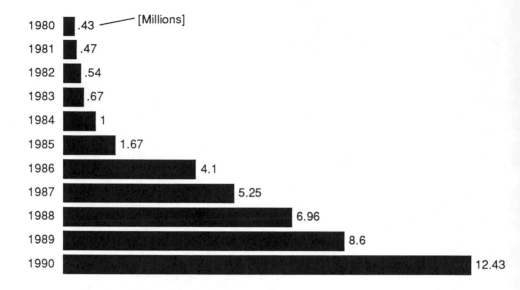

Figure 4.1 Public and private e-mail usage, 1980–1990. (Source: *Telecommunications*, August 1991, p. 15.)

other computer-mediated conferencing (CMC) systems, however, have tended to thrive in progressive corporate organizations, especially those that have the following characteristics:

- Decentralized and matrix-oriented management structures;
- Heavy and consistent emphasis on intracorporate use of desktop computing;
- Encouragement of individual initiative in both vertical and horizontal communications.

Digital Equipment Corporation, cited as one of the "100 best companies" to work for in the United States, was an early adopter of e-mail and other advanced internal communications techniques. DEC's decentralization, loosely structured management style, and emphasis on technical and engineering problem solving were major contributing factors to their successful implementation. DEC has also traditionally been a company that places a major emphasis on using its own products throughout its many facilities in the United States and Europe. It has also developed an admirable model of a private, globally deployed, enterprisewide network, perhaps one of the largest of its kind in use today. Over 110,000 DEC employees have access to the electronic mail system around the globe. Annual volume exceeds 42 million messages [1].

E-mail, when used in the right context and in the right environment, has the capacity to heighten both management and employee awareness within the corporation. Such awareness can, among other things, contribute toward the elimination of wasteful and duplicative procedures, the streamlining of general business practices, the exposure of suboptimal management approaches that have been allowed (for political considerations) to override wider corporate objectives, and the dissolution of traditional bureaucratic barriers created by unnecessary and unproductive interdepartmental competition.

Along with other electronic forms of communications, e-mail has been researched and studied with respect to these kinds of effects on intraorganizational communications. One study compared standard interpersonal and electronic communications patterns revolving around the specific context of business meetings [2]. The results were rather interesting. The study revealed a marked difference between the quality of the decision-making processes involved in both types of meetings. More specifically, it found that in electronic meetings, there was a far greater tendency for a range of group members to participate. By contrast, in face-to-face meetings, there was an increased likelihood for the group to form a consensus and mold their decisions on the basis of cues taken from the highest-ranking management representative present. This tendency of electronic communications to be more "democratic" in nature has interesting implications for how business might be conducted in the future. These implications extend to the manner in which employees might interact with both management and each other in group-influenced situations.

With this in mind, it seems fair to ask whether the egalitarian nature of electronic communications can provide internally driven competitive advantage with respect to corporate goals and objectives. One way to try to answer this question is to look at the effect of communications on organizational knowledge gaps, or areas in which traditionally poor communication has allowed suboptimal business practices to become ingrained and entrenched. Such gaps often develop when organizational politics stymies the natural flow of communications across traditional channels. Thus, the use and encouragement of more horizontal communications (such as that afforded by e-mail) can serve to ameliorate poor communication by providing alternate channels. Such approaches can help a corporate department or a small-to-medium-sized business operate more effectively, with positive impacts on customer service, satisfaction, and, ultimately, bottom-line profitability. Furthermore, electronic communication has the potential to make a group-oriented decision-making process less politically oriented and more task-oriented (that is, devoted more to solving the problem and less to maintaining the facade of good performance at the expense of same). Any increase in task orientation ultimately creates benefits at every level of the corporation, allowing departments and other functional units to compete more effectively in a world market that has come to value speed, responsiveness to the customer, and the minimization of bureaucratic entanglements that get in the way of delivering these benefits.

At the very least, e-mail is capable of bringing a much more collaborative dimension into business situations by introducing workflow principles that have been utilized extensively in other domains, most notably in the research and academic environment. For example, using the Internet, the massive, government-originated "network of networks," researchers and academics in many of the nation's top universities have pioneered the use of electronic information exchange, via e-mail. The Internet has for many years allowed academics to share research, comment on the progress of research efforts, collaborate on proposals, and share project-related information. It has become an indispensable means of communication, especially in the scientific community. The use of e-mail via the Internet and its associated regional and campus networks is now a well-established and common phenomenon throughout many of the nation's colleges and universities.

In the corporate domain, however, e-mail continues to experience an uphill climb toward acceptance, for the reasons cited earlier as well as other factors, mainly cultural in nature. When examining and understanding these differences, it helps to look at both formal and informal workflow procedures commonly in use throughout academia. Academics, as a result of both training and common practice, are fully accustomed to information sharing as a means of conducting their professional activities, thus enhancing both their own and the common knowledge base. In addition, they have learned to become adept at the process of "horse-trading" information far more effectively than their corporate counterparts. There is an implicit *quid pro quo* in passing along relevant information to a colleague in a comparable or related field, and the more competent electronic networkers soon learn that when information is conveyed, there is often an informal understanding concerning reciprocity. This is the way that informal but highly valuable information networks are developed among users.

By contrast, information in the corporate domain is viewed as a corporate asset and is, therefore, far more tightly controlled. Corporate management has traditionally played the information game in a completely different fashion, as dictated by the demands of expediency, the legitimate need for strategic secrecy, and the very real demands of corporate politics. In the traditional hierarchically organized company, the flow of information is generally more vertical than horizontal. The drawback is obvious: the more levels of management that exist in a given organization, the more bottlenecks for information flow that can develop, and—according to the assumptions of traditional communications paradigms such as the Shannon model—the more opportunities for distortion. This kind of corporate structure, far from fostering a free and easy exchange of information, serves to discourage that exchange. There are powerful political forces at work in this process, and, for middle management, maintaining control over the flow of information goes hand in hand with preserving political and managerial control over the day-to-day conduct of activities and operations under its purview.

Another organizational dynamic that militates against horizontal communication in traditional organizations is the need for secrecy with respect to maintaining strategic advantage. For obvious and understandable reasons, many companies are reluctant to engage in prematurely publicizing their plans and activities too widely, lest their competitors become the beneficiaries of early and gratuitous notice about new product and service strategies and directions.

COMPUTER CONFERENCING SYSTEMS AND BBSs

Closely related to e-mail is another electronic form of communications known as computer conferencing systems (CCSs). These are software-based systems that allow multiple parties in different geographical locations to participate in round-robin discussions. In these sessions, responses are posted in a sequenced, time- and date-stamped, serial progression. Such systems include options for moving responses or groups of responses to other sections of the conferencing system, deleting responses once they've been made, performing key word searches or author searches, and a variety of other features. This particular form of communication, perhaps even to a greater extent than e-mail, has had a difficult time breaking into even marginal corporate use, although it has found a measure of success in certain niche applications, as will be described.

Computer conferencing was originally developed by Murray Turoff in 1971. Turoff's system was called the Electronic Information Exchange System (EIES). Other CCSs include Picospan (a UNIX-based system developed by Marcus Watts), Participate, Caucus, and Notepad. Many of the developers of these systems hoped that they would become widely adopted in the corporate environment. Unfortunately, that did not prove to be the case, as described by Matthew Rapaport:

> In 1982, researchers like Dr. Turoff and Roxanne Star Hiltz were confident that computer-mediated communications systems would soon be embraced by corporate institutions worldwide. Their potential for reduced corporate travel, enhanced productivity through better communications, and better project management control (among other benefits) seemed too compelling to ignore. Yet as we move fully into the last decade of the 20th century, corporate interest in these systems is almost nonexistent. Aside from VaxNotes, whose use cannot even be estimated because it is bundled with other DEC products, there are less than 350 group-oriented communications systems employed by U.S. corporations for internal communications. The only bright spots in this otherwise bleak market picture are the growing use of small, PC-based

BBS systems for customer support services in technology companies, and the explosive growth of a nationwide public access network of electronic bulletin boards and large-scale communications services like H&R Block's Compuserve, Byte magazine's BIX, and General Electric's GEnie. [3]

As in the case of e-mail, the reasons for the slow acceptance of corporate CCS use are frequently cultural and political. In addition to these factors, others include the lack of user-friendly, interoperable graphics capability; the relative unavailability of such systems through commercial distribution channels (and the corresponding lack of marketing push that related products such as groupware enjoy); and the general reluctance of users to participate in systems not strongly endorsed and used by management.

In conjunction with CCSs, there are a variety of PC-based bulletin-board services (BBSs) becoming increasingly available for corporate use. These systems require the use of a dedicated PC that can act as a server to the participants using the system. The BBS can offer essentially the same conferencing features that are available on CCSs, but with the additional benefit of a higher level of user control. By contrast, traditional CCSs were more often designed for use with mainframes and minicomputers or, in some cases, were offered as a remotely accessed service.

What are some of the actual and theoretical uses for such systems? Customer support is one application that many companies are currently exploring. For example, instead of having customers call an 800-number for technical support, a BBS might be used in much the same way but would offer several additional advantages. First, when a customer sends in a question or problem using a text-based BBS, there is no initial waiting period until a service technician becomes available. Second, the complaint is documented, which gives the customer a record of the transaction and whatever responses were (or were not) forthcoming at the time. This can be used as documentation should any future problems arise and also fosters accountability on the part of the support staff. Third, the system provides both management and the support function itself with a continuous activity log of customer transactions. This log can be used to create an ongoing data base of problems and solutions, which can be accessed in reference to future interactions by both customers and support staff.

What changes will have to occur for these kinds of text-based conferencing systems to become more widely used in the corporate domain? For one thing, adoption and marketing push by a major computer vendor would certainly help. In addition, increased levels of interoperability between different users with different systems would further the cause. But it is also entirely possible that the larger window of opportunity for such systems has already been lost. This is not necessarily a drawback, but rather only a slight setback because the capabilities involved will likely become incorporated into the next generation of new work-group-oriented software systems (that is, groupware). In this sense, CCSs might even be considered as only the primitive forerunners of far more sophisticated systems yet to come.

In general, the rate of acceptance of such systems will depend on the development of other technological advances. For example, the interoperability of LANs is a critical factor in the ability of independent user software programs to intercommunicate. LAN internetworking became a major issue for corporations in the 1990s and will continue to grow in importance. For any type of primarily text-based electronic communications systems to achieve the requisite ability to garner more serious support among senior managers, such systems will have to be broadly interoperable among many groups throughout the corporation, including its geographically dispersed facilities.

Some groupware systems have spread throughout corporate departments primarily by word of mouth. As a result, various groups of employees have begun the independent process of attaching to such systems simply on the basis of their own enthusiasm rather than as the result of any managerial prompting. Such spontaneous and unsolicited usage is of course the best way for these technologies to gain acceptance and to make their mark within the corporate rank and file. Certainly the success of various PC-based programs was due in large measure to the tendency of one user to recommend it to another.

Another key issue besides interoperability is bandwidth capacity. Users who are sharing files and documents must be able to transparently pass this information among themselves in such a way that it appears that they are using their own PCs in the normal fashion and not accessing anyone else's files at a distance. Closely related to this is the need to share not only text but graphics and compound documents as well. Thus, not only will interoperability among hardware systems be required, but also compatibility between the command structures and user interfaces found in the products provided by multiple vendors.

COMMERCIALIZED USE OF THE INTERNET

The Internet is a large and complex TCP/IP-based research and academic "network of networks" that connects over four million users in the United States and at least thirty other countries. There are approximately 5000 individual networks on the Internet and 500,000 connected computers. Configured in a complex tangle of campus, regional, midlevel, and backbone networks, it was originally sponsored and funded by the U.S. government as ARPANET and primarily existed for use among various defense-related government agencies and their contractors. Currently, certain U.S. portions of the Internet are undergoing a major metamorphosis as they evolve into the National Research and Education Network (NREN).

Because the Internet was originally intended for use as a research and academic network, commercial traffic has traditionally been disallowed on the core network—the Internet backbone—under an acceptable-use policy developed by the National Science Foundation (NSF). (NSF is the federal agency responsible for the Internet's operation and general management.) However, in 1987, NSF put out a competitive bid for the physical management of the Internet backbone, which was granted to a nonprofit organization at the University of Michigan called Merit. Merit developed cooperative agreements with other entities for the actual operation of the network, including MCI for basic transport and IBM for general technical support and capability on the computing side. The backbone has been operated by a nonprofit venture called Advanced Network Services (ANS), a collaboration of IBM, MCI, and Merit Networks, formed in 1990.

In the late 1980s, demand for commercial use of the Internet increased, and several firms—including Performance Systems International (PSI), CERFnet, and UUNET Technologies—began marketing Internet services on a commercial basis. In 1991, these three service providers moved to establish the Commercial Internet Exchange (CIX), which allowed them to exchange traffic among their respective networks. This provided yet another boost to Internet commercialization. In 1992, Sprint joined the CIX member companies and began providing commercial Internet services primarily to its existing government customers under the FTS-2000 contract. Over 3000 firms are now connected to the CIX, including the top twenty computer companies in the United States.

In June 1991, a major change in the Internet landscape occurred when ANS began to market commercial services on the Internet directly to corporate customers without any acceptable-use restrictions. This was accomplished via a spinoff-for-profit subsidiary chartered to develop the T3-based Internet backbone and market it for commercial use. The subsidiary—ANS CO + RE Systems, Inc.—has added high-end corporate users in the Fortune 1000 to its overall customer base. The commercialized service represents a vehicle for allowing corporate and research/education users to interconnect as well as a means for companies to interconnect their own geographically dispersed networks. ANS's intention is to continue to develop a national data network utility that can offer high-bandwidth connections to corporate customers, and extensive national coverage. For example, high-bandwidth migration to the OC levels of the SONET hierarchy is being planned. This commercial facility, called ANSnet, is a virtual network that overlays the Internet backbone and can be used as an alternative to a nationally deployed private T1 network, for example. Furthermore, an important potential benefit for corporate users is the fact that the current networking capabilities of many companies are already based on the TCP/IP protocol suite—the same internetworking protocol used by Internet. In addition, with the installation of more and more desktop computer systems based on the UNIX operating system, which has inherent and built-in TCP/IP affinities, the desirability of TCP/IP-based networking should also increase.

In addition to the commercial services being provided by ANS, the CIX providers represent another major alternative for corporate users. Moreover, the individual attached midlevel networks that operate off of the Internet backbone—such as NYSERnet and SURAnet—are also increasingly providing commercial service on their own networks. The entire commercialization issue has been plagued with controversy and issues of fair network access, because, at least in the academic community, Internet access has been virtually free to qualified users. Such issues are being resolved between ANS, which critics charge has an unfair advantage with its preexisting NSF relationship, and the CIX, which says that the market needs to be opened up for more competition to create a level playing field [4].

These complex regulatory and political issues aside, how can corporate users benefit from Internet services? As it stands now, the Internet can be used by corporations for a variety of purposes, including

- Pure transport capability at speeds ranging from dialup to T3;
- Access to major research and academic institutions;
- Access to individual regional or midlevel networks;
- Electronic mail and file transfer;
- Access to newsgroups and other current subject area discussions;
- Access to independent information providers on the net, such as Dialog.

Many large companies, such as DEC, have been on the Internet for years as a result of their heavy contractual involvement with government projects. Within Digital, for example, product research and development teams use the Internet to communicate with researchers in the academic community. Such use, however, has been restricted to the previously mentioned policy of acceptable use. Hewlett-Packard is in a similar position. Like DEC, HP has used the Internet for applications related to research and academic communications. The Internet-connected regional networks that HP was initially tied into included Colorado Supernet, BARRnet, and CREN. However, to extend their use of the Internet for more strictly commercial purposes, they signed an agreement in 1991 with PSI, one of the original CIX partners. HP now uses the service for three basic applications:

1. The exchange of design files;
2. Software upgrades for existing products installed in the customer base;
3. Giving customers access to data bases addressing HP products and services.

The growth of Internet, in terms of both nodes on the network and traffic accommodated has been nothing short of phenomenal. There have been dramatic increases in both commercial use and the actual number of subscribers signed on. Moreover, the network is undergoing increased international expansion. As use continues to grow and bandwidth capabilities are expanded, corporate users will likely find many more applications available to them. There is, for example, much speculation that the commercialized Internet could provide a robust platform for a growing number of advanced information services. Proponents of the NREN argue that, with this kind of capability distributed throughout a wide subscriber base in the educational community, the network could become an extremely versatile tool for information exchange and learning. Dialog, the well-known data base provider, is already on the Internet, and others are sure to follow. In addition, there are new search engines and technologies being developed by companies such as Thinking Machines that will allow the Internet to become far more user-friendly in terms of the simple process of sifting through the massive amounts of information that will be available on it. The Thinking Machines concept is known as a wide-area information server (WAIS), which is already being tried for some Internet applications. Another important leading-edge technology is being explored by the Corporation for National Research Initiatives (CNRI), an R&D entity that fosters Internet growth and development with funding from both NSF and DARPA. The CNRI project involves another type of intelligent search engine called Knowbots, an application designed to rove through the network and automatically search material based on user-selected descriptions and parameters.

Corporations that are exploring the commercial use of the Internet, however, face a steep learning curve. Many aspects of the system taken for granted among net-savvy academicians are virtually unknown to corporate managers in the Fortune 1000. The key toward further progress, however, can be summed up in one word: exploration. Companies that allow their employees the latitude to explore this increasingly important resource will find that the Internet-literate user is able to access a wide variety of rich information sources. Once these sources find their way into the company as the result of the efforts of enterprising network "information surfers," the way becomes paved for incorporation and consolidation of this information as an internal corporate resource that can contribute to overall marketplace competitiveness.

Finally, there is the phenomenon discussed in Chapters 1, 3, and 7 involving perhaps one of the most intriguing aspects of Internet: using it a means of creating new "virtual ventures" and *ad hoc* enterprises involving a variety of entities in both public and private sectors. At a time when the so-called military-industrial complex is being downscaled and downsized, the network that it spawned—the Internet—can be viewed as a means of channeling otherwise-impossible synergies between the defense community, the research/academic community (much of which is engaged in defense-related activities), and corporate development of products and services. Given the fact that the ability of the United States to compete in a newly emerging global marketplace is being reexamined, such a means of information and technology transfer might allow the United States to compete more effectively by creating new models of cooperation between the three sectors. An excellent example of such a likely future scenario (described in Chapter 7) is how an Internet midlevel network called SURAnet was used as the basis for developing a regional network in the southeast that accomplishes this very objective. The next-generation Internet/NREN could, with the right kind of developmental guidance, provide the enabling technology to act as a catalyst in the creation of new modalities for cooperative and collaborative arrangements in the American workplace.

ELECTRONIC INFORMATION SERVICES

Electronic information services are an important part of a company's internal resource base in the ever-changing landscape of the information age. In the not-too-distant future, they may come to be viewed as one of a company's most important resources for competing in the new information economies that are now developing. Such services, including CompuServe, LEXIS/NEXIS, and NewsNet, have been in use for some time, albeit in a limited fashion and primarily for niche applications (see Table 4.1). Many such services were deployed gradually throughout the 1980s but were procured on a departmental basis, much the same as LANs grew in scope and number during the same period. Although electronic information services still lag behind LANs in the degree to which they are fully engaged in the corporate setting, network managers should assume that they might eventually inherit administrative oversight for these corporate resources.

LEXIS and NEXIS are remote data base services marketed to commercial and corporate customers for specific sets of applications. Users who sign up for the service are provided with an on-site terminal and charged for the specific services they select on a usage-per-minute basis. The LEXIS service is targeted as a research tool for the legal profession. Its companion service, NEXIS, is a more generalized data base and includes such items as daily transcripts from a number of major U.S. newspapers; on-line versions of trade magazines, journals, and other periodicals;

Table 4.1
A Sampling of Information Service Providers

Provider	Service	Category
CompuServe	CompuServe	general interest
Mead Data Central	LEXIS/NEXIS	professional/technical
NewsNet	NewsNet	news service
US West	CommunityLink	RBOC gateway
Bell Canada	Alex	RBOC gateway
IBM/Sears	Prodigy	general interest
Dow Jones	DJ News/Retrieval	financial
GE Info Services	GEnie	general interest
US Videotel	US Videotel	general interest
Dialog	Dialog	professional/technical
AMIX	AMIX	business transactions

Source: Information Industry Bulletin.

and other news and professional source materials. The on-site terminal provided also allows a user to perform key word searches of the data base and capture it either electronically or to hard copy.

Some of these data bases have been in existence for a number of years but were available only under the auspices of a company librarian. However, in keeping with an across-the-board tendency toward decentralization, such services are becoming available directly to end users throughout the corporation. Furthermore, moving such capabilities out to the end user is of critical importance: sitting at a PC and instantaneously being able to call up a rich palette of information services is a far cry from having to walk two buildings away, fill out a search request form, obtain departmental authorization, and get the results of a data base search back in one or two weeks. Thus, *accessibility* to information services, made possible by PCs, PC LANs, and the requisite local- and wide-area connectivity are critical dependencies in the strategic use of information. Unfortunately, it will be some time before such capabilities are routinely incorporated into the array of professional resources normally made available to users. There are many reasons for this, but first and foremost is cost, and the attendant difficulty in justifying this kind of availability as a valuable investment.

Other providers of information services include the RBOCs, which, via the loosening of regulatory restraints, are now allowed to provide content-based services, whereas they had been previously limited to trials of information gateways for other providers. A number of the RBOCs were at one time involved in the latter endeavor. One highly publicized venture into this market was Southwestern Bell's Sourceline. Other RBOCs involved in trials include Nynex's Info-look service, U.S. West's Community Link, BellSouth's TUG, and Bell Atlantic. Many of the RBOCs now have full-fledged content-based information services in place, and some, such as Nynex, are offering electronic yellow pages. In areas outside the United States, France Telecom's experiment with Minitel has been widely publicized and discussed. (It's interesting to note that several RBOCs—including U.S. West, Bell Atlantic, Nynex, and BellSouth—have been running Minitel Services over their gateways.) Another important effort is Prodigy, the IBM/Sears partnership, which has upwards of a million subscribers.

Information services are finally beginning to achieve critical mass as they find their markets. The industry has managed to decisively shake the industry image of a technology begging for customers, promulgated in part by the earlier, costly failures of ventures like Knight-Ridder's Viewtron and Times-Mirror's Gateway. There are several keys to the current successes. First, cost factors are now much more in line with what consumers are willing to pay for such services, in part thanks to Prodigy's success with flat-rate pricing (rather than per-usage charges). Second, PC penetration in the United States is now estimated at 15 to 20% of all households, and rising.

Another well-known and widely used service that falls into the category of electronic information services is CompuServe. CompuServe, which was first available in 1979, was one of the first services to be offered both to businesses and the general public. Like Prodigy, it has over a million subscribers and is widely used throughout the business community and by individual subscribers. CompuServe is based on an annual membership fee and usage-sensitive pricing (hourly charges), on the order of $11 to $13 per hour. In 1991, the service began to offer a flat-rate pricing option similar to the approach taken by Prodigy and GEnie, GE's on-line information service. This option, however, also imposed limitations on the full range of services that are available through the regular subscription rate.

Some of the services available over CompuServe include

- News and information, including summaries from the Associated Press, UPI, Reuters, and major newspapers such as the Washington Post. In addition, CompuServe offers a "clipping service" that monitors news in a user-selected subject area. For example, the service might forward all news items on a daily basis related to the subject area "chemical manufacturing."
- E-mail, which allows all CompuServe members to communicate among themselves privately, as well as with users in other e-mail systems such as MCI Mail.
- Online data bases in the areas of science, medicine, law, and a variety of other technical and professional disciplines. In addition, access is available to professional journals, specialized newsletters, and published research papers. Demographic data is available for companies interested in selecting business locations, performing competitive analyses, or developing direct-mail marketing campaigns.
- Financial and business information, such as stock market activity reports and reference sources such as Standard and Poor, Value Line, and Disclosure. Users can access, for example, up to twelve years of weekly and monthly pricing histories for various investments.
- An "executive service option" to accommodate the needs and objectives of executives, including custom design of news and information inputs from a variety of standard news sources such as AP, UPI, and the financial and business wires; business decision support services, including market research reports; and databases.
- Hardware and software support conferences in which users can resolve problems via communication with experts familiar with their systems. Technical representatives are available from such vendors as Ashton-Tate, Adobe, Aldus, Xerox, Novell, Lotus, and Microsoft. A number of companies have selected CompuServe as their official technical support system.
- Special-interest group (SIG) forums, in which users gather to discuss a wide

range of subjects, many of which are related to professional and business activities.

- Travel services, including access to the SABRE system for airline reservations. The service allows users to comparison-shop fares and book on-line reservations.
- On-line purchasing, including access to retail stores, specialty shops, and discount wholesalers.

Many of these services and options can be used to support a wide range of corporate business activities. For example, purchasing departments might utilize the on-line purchasing feature to establish pricing baselines. A department experiencing problems with their internally purchased PC or LAN equipment might use CompuServe vendor conferences as a source of technical support apart from what might be available from in-house MIS services. (In fact, MIS might avail itself of on-line access to representatives from Lotus or other software houses.) Marketing professionals or strategic planners might use the special-interest groups to participate in discussions about future business trends and directions. Corporate travel departments or departmental support staff might use the travel service as an alternative to travel agencies. Managers and executives in a variety of positions could benefit from routine access to financial, global, and business news, especially when filtered through a "clipping service" option, which scans for relevant key words or other information indicators. Access to e-mail has already been covered extensively in this chapter, although it's worth pointing out that the e-mail function in the context of a service such as CompuServe or Prodigy is almost a value-added benefit. When coupled with access to other e-mail addressing destinations available via X.400 or other connections to different services, e-mail access can become a powerful networking tool. Finally, it goes without saying that the ability to access on-line data bases of varying descriptions is a capability that can prove useful for many business applications.

Should such services be more widely deployed in companies? What would be the expected short- and long-term paybacks? Would making them more widely available tend to encourage use for the wrong purposes? Undoubtedly, the common conventional management wisdom holds that it is extremely difficult to cost-justify such services, especially given the high hourly rates associated with many of them. Other legitimate management concerns include the prospect of employees using aspects of the service that aren't specifically relevant to their job functions. One scenario that might serve to alleviate such concerns would be to make it available with specific service restrictions, an unbundled approach likely to become far more common in the future.

As for cost justification, things get trickier indeed. It's difficult to quantify the enhancement of an employee's knowledge base or the return on investment for keeping up-to-date in a given field or functional area, as any director of employee training knows. However, in the new information-based global marketplace, keeping current on professional matters will become increasingly critical, especially because the rate of change of technological development and applications will only continue to accelerate. Thus, conventional thinking must be reexamined in terms of the need to keep staff up-to-date on developments directly or indirectly affecting their professional sphere of influence. Whereas it is always difficult to know what a staff member might be doing while sitting in front of a PC, the fact of the matter is that it is the users themselves that know best how a particular service can help them optimize their own job function. No amount of micromanaging is likely to change that situation if a staff member can at least be considered a reasonably responsible member of the team. Corporate managers, therefore, should remain as open-minded as possible regarding the appropriate use of these services.

CRITICAL TRENDS IN INFORMATION SERVICES

Due to the the the time-sensitive nature of information in a globally networked electronic environment, it seems clear that competitive advantage can be gained from the use of advanced information services. When events are constantly being propelled into greater visibility and rapidity of occurrence via electronic communications, those without the ability to readily access the near-continuous stream of available market-related information will be at a distinct disadvantage to those who do have such access and know how to use it intelligently. This being the case, companies would do well to not restrict the flow of information services just to the managerial level. As I hope to demonstrate, the nature of information networking is such that accessibility at all levels is a critical dependency in using information internally to foster competitiveness.

At a time when an exponentially increasing base of information is being created and disseminated more rapidly than our ability to process it, questions arise concerning the tools and technologies that might be available to help professionals filter and pinpoint vital information awash in a Sargasso sea of undifferentiated data. Such tools do exist, although we are a long way from enjoying the kind of sophisticated capabilities that approaches such as WAIS and Knowbots, discussed earlier, might someday be able to provide. Thus, tailoring information in its raw form for specific business purposes is rapidly becoming an indispensable means of self-defense for more advanced information gatherers in the corporate setting. The ability to use techniques to customize an information stream keeps a user from being overwhelmed with extraneous information that can negatively affect the productivity it was intended to foster, as well as avoiding the highly undesirable and all-too-common condition known as information overload.

What are the tools that now exist for dealing with this very real problem—the downside, if you will, of the information revolution? As mentioned previously, many electronic information services such as CompuServe have custom-tailoring options that can serve to direct specific kinds of information to the user and filter out unwanted data streams. Some voice processing companies such as Octel offer services that accomplish the same objective using voice messaging rather than data as the means of delivery. However, these kinds of information filtering approaches are still in their infancy. Eventually, aided by such technologies as artificial intelligence and hypertext, new tools will appear on the commercial horizon that will tailor and channel information toward the user, as opposed to its undesirable opposite: immersing the user in a morass of information that lacks the proper navigational guideposts.

Thus, the question of exactly how electronic information services can best be used as an internal strategic tool remains to be answered. At this point, there are more questions than answers concerning how such services will be used in the corporate context and how extensively they will be deployed. Such questions include: How can these services be used to enhance areas such as the training and education of corporate staff? What role will they play in the reinvented service economies of the 1990s? And, most significantly, to what extent will they become models for new forms of information production and consumption, with significant implications for how corporate users routinely conduct their business?

Let me try to illustrate more graphically what I mean. If you examine the trend toward the deployment of information services in the context of some other major trends discussed earlier—vertical disintegration, for example—some interesting possibilities emerge. Seen in this light, information services not only can be viewed as a means of enhancing the work of symbolic analysts—corporate professionals—but also can form the basis for fundamentally new approaches to the way that business is conducted. For example, companies in the service and retail sectors are increasingly relying on access to their customer base via on-line networking. Both Prodigy and CompuServe have experimented with providing shopping services as well as professional services via their systems. Prodigy, in fact, even ran some trials with supermarket chains, whereby the service could be used for home delivery of groceries. Just as the rise of overnight delivery services prefigured the advent of facsimile, the current growth in catalog-based mail-order business might prefigure the move toward direct company-to-customer on-line business transactions. When either the number of households with PCs in the United States begins to approach 30% or the number of communications-capable PC households approaches 50%, it's likely that many retailers will begin to mount major initiatives toward marketing their products and services directly to households.

Another trend that relates to the increased use of these services is the reinventing of the employer-employee relationship, discussed in Chapter 3. In the not-too-distant future, home-based professionals may come to function more and more as independent consultants. Through the use of advanced information services, individuals with particular skill sets (accountants, title searchers, technical consultants, for example) would be able to set up shop electronically and reach an extended and even national client base via the capabilities of networking. (On a relatively minor scale, this has already happened via the use of BBSs.) By the end of the decade, it may be common to see professionals working full forty-hour weeks at home, but for two or three major corporations. Such arrangements can provide increased flexibility with resource efficiencies being generated for both employer and employee. Such individuals might also develop long- or short-term *ad hoc* associations (that is, small temporary businesses) electronically with other professionals in their area of expertise. Furthermore, as new electronic services and niche markets are created, each of these will likely, in turn, generate a new constellation of professional/symbolic analyst types of positions based on the new capabilities they might provide. Thus, for example, an enterprising information provider might begin to gather statistics about various communities associated with major metropolitan areas in the United States. In turn, a subscriber to that service—another entrepreneur—might then take that information and create a position selling those statistics in repackaged, user-friendly form to prospective home buyers. The information value chain that's created will thus generate its own, often self-replicating, opportunities.

Finally, moving beyond even the retail market, information services, once widely deployed throughout U.S. homes and businesses, might also act as clearinghouses for many types of business transactions. Realtors are already showing homes to prospective buyers electronically. Architects are exploring the use of virtual reality to custom-design complex buildings. Electronic information services can increasingly be used to broker service transactions between corporations and their own suppliers and service providers. For example, in France, the Minitel system is used to match trucking services to companies that have materials slated for shipment. Once data networking achieves the same kind of critical mass that voice services now enjoy under the concept of Universal Telephone Service, the information gateways of the future will become virtual marketplaces for a wide variety of corporate, commercial, governmental, and personal transactions.

GROUPWARE: THE FUTURE OF THE OFFICE

One of the most interesting and yet largely unexplored areas of internal communications that can contribute to strategically competitive performance is a somewhat nebulous and still emerging technology known as groupware. What exactly is groupware? Technically, it is a computer-based system that is primarily used to facilitate task-oriented electronic communications among workgroups or teams. However, there are many systems that seem to fall into this category but aren't really groupware, including traditional office automation devices and even the telephone. However, until newly emerging types of groupware themselves become more well-known and definable, we may have to live with a certain amount of conceptual fuzziness.

A little history might help. According to one source, the term groupware was coined by Peter and Trudy Johnson-Lenz, who first wrote about it in the early 1980s [5]. The roots of its invention can be traced to Stanford Research Institute's Douglas Englebart, also known as the inventor of the computer mouse. Englebart and his researchers were looking into new ways to improve the human working interface between computers and users. In the course of these efforts, they developed a cooperative teaming computer system called Augment. Augment, one of the first known groupware systems, combined the functionality of hypertext with that of more or less traditional computer conferencing systems. In fact, most early forms of groupware fell more or less into this latter category, although eventually their paths would diverge, and groupware would become quite distinct as a software product.

During the 1970s, the ARPANET began to incorporate conferencing capabilities, along with the e-mail functions that were an inherent part of it. This was the beginning of what are now known as Internet news groups, wherein topics on a wide variety of subjects allow members to comment in chronological succession. Around this time, there was also a lot of work being done at the New Jersey Institute of Technology, where the Electronic Information Exchange System (EIES) was developed by Murray Turoff and Starr Roxanne Hiltz. These early conferencing systems were primarily targeted toward the academic community, and to this day it is this community that is far more familiar (as were early ARPANET users) with the important work-teaming capabilities available through such systems.

One way of distinguishing groupware from some associated technologies is to look at the basic patterns of communications involved in each. E-mail, for example, is part and parcel of any groupware system and shares certain characteristics with it. However, e-mail is classified as "one-to-one" communication. Similarly, BBSs also share many characteristics of both groupware and e-mail systems but are generally limited to the "one-to-many" communications paradigm. However, both groupware and its cousin, CCSs, are distinguishable from the others as falling into the category of "many-to-many" communications.

Opper and Fersko-Weiss delineate four categories of groupware functionality, based on the applications involved in each [6]. These are administration, information management, communications management, and real-time meeting facilitation.

Administration-oriented products are targeted toward such functions as calendaring and automatic scheduling of meetings. They allow communications to take place in environments in which nonoverlapping personal schedules militate against effective departmental coordination. *Information management systems* are geared toward activities such as joint document authoring, project tracking, and critical path management of the sort traditionally done manually via the use of

PERT or Gantt charts. *Communications management* provides platforms for "virtual" meetings among team members. This type of groupware is closer in nature to the CCSs mentioned earlier. *Real-time meeting facilitation* is perhaps the newest, most innovative, and least well known of the four categories. These systems are used not simply to schedule meetings or provide a virtual electronic shared space for them to take place; rather, they are primarily intended to enhance and improve the quality of both face-to-face and electronic communications by providing validation, feedback, and machine-based objectivity—qualities that are often conspicuously absent from traditional meetings as a result of the human propensity toward "political" solutions, which often keep participants mollified but fail to solve the problem at hand. Two fairly sophisticated types of systems in this category are voting systems that tabulate individual selections of meeting participants working on common problems and scenarios and systems that give each participant a computer during a meeting, which is then used similarly to help arrive at common solutions and areas of agreement.

It is highly likely that by the middle of the decade, a significant portion of business activity will be organized around the groupware concept, especially given its propensity for facilitating interaction that is time- and distance-insensitive. This trend will be driven by other developments in the working environment, including the movement toward flextime and staggered working hours, increased utilization of telecommuting options by corporations and their employees, and the increased tendency on the part of companies to outsource certain functions to independent contractors. The benefits in productivity derived from such approaches are potentially quite significant.

No discussion of groupware and its applications would be complete without also discussing one of the first—and most widely known—commercial offerings associated with this technology: Lotus Notes. One market research firm, Forrester Research, has already predicted that products such as Lotus Notes will be adopted largely as the result of a groundswell of popular support and demand from users who find that it increases their independence, involvement, and decision-making latitude and frees them from the politically driven dependencies of traditional approaches to office communications. Other important LAN-based groupware products are AT&T's Rhapsody, HP's New Wave, and NCR's Cooperation.

First available in 1989, Notes appears to have received very favorable response from users; whether it will become the 1-2-3 of groupware products remains to be seen. The Lotus Corporation describes their product as capable of providing such applications as "customer tracking, status reporting, project management, information distribution, electronic mail and collaborative free-form discussions of all kinds" [7]. Figure 4.2 shows where Notes is positioned in the array of other groupware and groupware-related products on the market.

The foundation for Notes is a shared data base, represented as a series of six file folders, used as real-time workspaces. Each folder contains user-designated icons that identify which of the various groupware applications described above

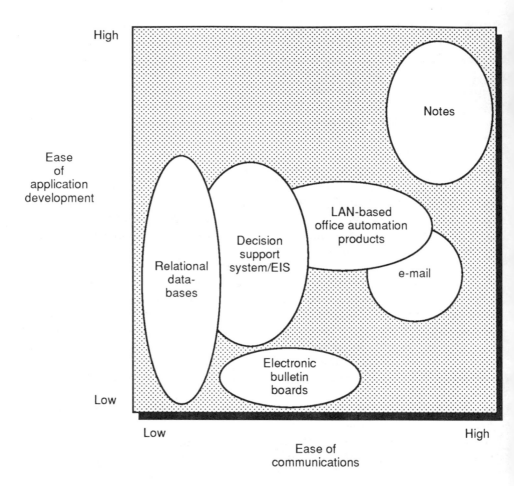

High

Ease
of
application
development

Low

Notes

LAN-based
office automation
products

Decision
support
system/EIS

Relational
data-
bases

e-mail

Electronic
bulletin
boards

Low   High

Ease of
communications

Figure 4.2 Lotus Notes: market position versus competing systems. (Source: Forrester Research.)

the user is currently configured for. The actual data base, however, not only supports pure text, but is multimedia in nature and can accommodate high-quality graphics and photographs. Architecturally, Notes is based on the client-server computing model but does not require continuous interconnection. It currently supports major operating systems, including DOS and OS/2, and the graphical user interface supports both OS/2 Presentation Manager and Microsoft Windows. Notes will run on any workstation that accommodates either of these two systems; servers, however, must be running on OS/2. In terms of communicating with other systems, the product supports most popular operating systems in use, including Novell NetWare and Banyan VINES.

Lotus Notes is most frequently used for four major applications:

1. Sales management;
2. Product development;
3. Customer service;
4. Executive information.

One company currently using the system is Compaq, which has approximately 2800 Notes licenses. An application in use at that company is a system called TechPAQ, which supplies technical information to customers about Compaq products and how to best integrate them with third-party network and multiuser operating systems. Other Notes users include MCI, Texaco, Metropolitan Life, and Nynex. At Nynex, the product is used to transfer engineering drawings, text, and scanned photographs between New England Telephone's Network Engineering Headquarters and other Nynex facilities. Figure 4.3 depicts how Notes is typically deployed in such environments.

A report from Forrester Research states, "Lotus has charged off into unknown and uncharted territory with Notes: the product simultaneously intrigues, excites, and bewilders the Fortune 1000. Notes converts are convinced that they have found a product which makes intelligent use of their computer resources. To the unini-

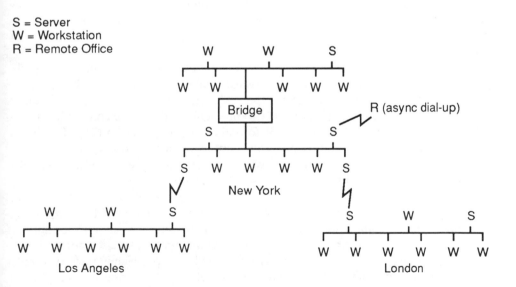

Figure 4.3 Example of Lotus Notes used in a wide-area environment. (Lotus Notes is a registered trademark of Lotus Development Corporation. Used with permission.)

tiated, Notes is a confusing and impossible to define product that violates all conventional product descriptions" [8]. However, the report also describes Notes as "threatening to MIS, in many case because they simply don't understand it and aren't familiar with such areas as OS/2, GUIs, and client server architectures." Even more interesting, it observes that Notes is frequently brought in to solve a tactical problem, but once a company has spent the $60,000 required for a one-time site license, the use of the product tends to grow rapidly among enthusiastic users, and it eventually becomes a strategic product for the company.

TELECOMMUTING AND THE NEW WORKSTYLE

There are many forces in the marketplace conspiring to change the traditional ways that employees interact with their employers. Many large corporations are now outsourcing tasks and subtasks to smaller companies or individuals outside of their organizations. Where the latter is involved, a contractual or consulting type of relationship can exist, even when the nature of the work performed is essentially the same as that which might be done by a full-time employee. One of the major benefits provided by such arrangements is increased flexibility for both parties, although there is no arguing that the economic downturn during the early 1990s was a major contributor to this trend. An important facet of this decentralization of corporate functions is telecommuting. Telecommuting can be defined as a workstyle that allows a corporation's employees to work at locations other than the company's corporate facilities (usually at home), as enabled by state-of-the-art computer networking systems.

Telecommuting is a trend that began to gather serious momentum in the early 1990s, although a relative handful of forward-thinking companies had been experimenting with it off and on throughout the 1980s. In addition to special and isolated cases of employees linked electronically to their employers, some companies incorporated telecommuting as part of their formal corporate policy. AT&T, for example, has a major telecommuting program underway on the West Coast. It began as a trial in July 1989, involving approximately 130 employees based in Southern California. Buoyed by its initial success, the program has since been officially adopted. Although AT&T employees frequently work at home via informal arrangements with supervisors, this was the first formal, measured telecommuting program that the company implemented. AT&T has since expanded the number of participants to over 300 and has other programs underway elsewhere. (Interestingly, in California, AT&T originally developed this approach in part to comply with state-mandated air pollution reduction regulations requiring that employers reduce their employees' commuting time.)

Estimates of the number of employees that will likely be telecommuting by the end of the decade vary considerably. Regional holding company BellSouth expects about 35% of their subscriber base (that is, the general workforce) to be working at home by the end of the decade, a striking number and one that BellSouth appears to be taking seriously in terms of its long-range strategic plans. An annual study performed by one market research company derived numerical estimates for the number of U.S. workers engaged in both work-at-home and telecommuting-based work-at-home [9]. The study found that the number of workers who are based at home during normal business hours (on either a full- or part-time basis) reached 5.5 million in 1991. This represented an increase of 38% over the 4 million counted in 1990 (see Figure 4.4). More significantly, the study found that telecommuting represented the "fastest growing segment of the work at home trend." It also found that the trend is growing especially rapidly in several key groups:

- Large organizations with over 100 employees;
- Very small organizations with under 10 employees;
- Business executives and managers;
- Engineers and scientists.

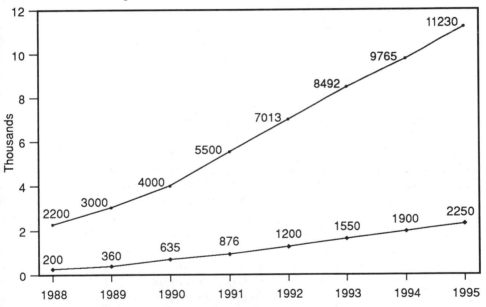

U. S. Telecommuting Outlook

At current growth rates, home-based telecommuting will double by 1995.

→ All US home-based telecommuters
→ Work at home 35 or more hours per week, Monday-Friday
Note: Telecommuters are defined as company employees who work at
home part- or full-time during normal business hours.

Figure 4.4 Projected trends for telecommuting. (Source: *Telecommunications*, September 1991, p. 13.)

How were various subgroups that were early adopters of the telecommuting trend categorized? To begin with, the study found that 43% of telecommuters comprised executives and what were termed "professional specialty occupations." On the other side of the coin, nearly a quarter of the sampled population were involved in what the study called "a variety of manual and low-tech jobs." These are assumed to be positions involving such activities as text-to-computer data entry, word processing, order processing, and so on. In further characterizing the nature of individuals generally involved in telecommuting, the study found that a typical telecommuter is either male or female, "thirty-something," and a member of a dual career household.

In addition, the study also looked at the technology behind these trends. For example, it found that computers were owned in 36% of telecommuter households as opposed to only 15% in U.S. households in which no homeworkers were present. The use of modems was tracked in 16% of telecommuting households; fax machines were involved in 7%. The study also found that spending by telecommuters on such equipment as PCs, software, peripherals, faxes, and telephone services totalled close to $1.6 billion in 1990, up 14% from $1.4 billion in 1989. Finally, the report predicted that the current base of telecommuters will double by 1995, totalling more than 11 million workers.

These trends have a number of interesting implications for corporations in transition. Telecommuting is, in essence, a winning proposition for both employer and employee if administered and structured correctly. The benefits for the employer include the establishment of a more flexible work force, the reduction in the amount of physical plant and facilities space required, increased access to employees during travel, reduction in the requirement for relocating key employees (and the attendant cost savings), greater availability of skilled professional talent (based in part on parents who prefer to stay at home), increased employee morale and option fulfillment, and increased employee productivity. The benefits to the employee, similarly, are also numerous. Employees often find that the flexibility afforded by such arrangements means that their relationship to their company evolves to that of consultant-client, rather than strictly employee-employer. In addition, telecommuting also can provide cost and time savings in terms of transportation, cost savings accruing from less dependence on day-care facilities, greater productivity resulting from the ability to structure the work day around personal requirements, and less stress in the work environment as a result of the decrease in organizational pressures. For these and other reasons (such as positive environmental impacts), telecommuting is likely to become widespread throughout the 1990s. In doing so, it will provide companies with greater productivity and flexibility in establishing a talent base that can keep payroll costs in line with the fluctuations of marketplace demands and other external variables.

NOTES AND REFERENCES

[1] For more on Digital's wordwide network, see Peter Brown, "Business Productivity Through Networking," *Telecommunications*, April 1992.

[2] Lee Sproull and Sara Kiesler, *Connections: New Ways of Working in the Networked Organization*, MIT Press, Cambridge, MA, 1991, Chapter 4.

[3] Matthew Rapaport, "Groupware vs. CCS: Comparing Benefits and Functionality," *Telecommunications*, November 1991, 37.

[4] For more information, see the following articles:
Vinton G. Cerf, "Another Reading of the NREN Legislation," *Telecommunications*, November 1991, 29–30.
Jay Habegger, "Why is the NREN Proposal So Complicated?" *Telecommunications*, November 1991, 21–26.
Thomas S. Valovic, "ANS and Internet Commercialization," *Telecommunications*, July 1991, 4.

[5] Susanna Opper and Henry Fersko-Weiss, *Technology for Teams*, Van Nostrand Reinhold, New York, 1992.

[6] *Ibid.*, 31

[7] *Lotus Technical Notes Series*, Vol. 2, 26 March 1991.

[8] George Colony and Stuart Woodring, "Lotus Notes," Forrester Research, Cambridge, MA, 1991.

[9] The study was conducted by New York–based market research firm Link Resources, as part of their sixth annual "National Work-at-Home Survey."

Chapter 5
Public versus Private Networks:
Organizational Dynamics

Two of the biggest questions organizations will face in the 1990s will
concern what investments they should make and what kinds of expertise
they should develop. That is, what part of the network do they want to
own and control, and what part do they want to buy from others? The
answers to these questions are inextricably entwined with a firm's overall
business strategy.

—William Johnson, Vice President,
Digital Equipment Corporation

How corporate organizations view options and possibilities for the strategic use of
communications depends on a variety of factors. One of the most critical factors
is the degree to which senior management in any given organization has decided
to handle the development of computer and communication resources as an internal
function, or whether the care and feeding of network resources is best handed off
to an outside organization. In the case of the former, there are the predictable
issues of requisite and expected investment in staff, resources, equipment, and
training to contend with. With the latter option—commonly called outsourcing—
there are many critical decision points that have to be carefully weighed before
moving ahead. Chapters 5 and 6 deal with this pivotal business dilemma. Chapter
5 focuses largely on the organizational developments that have occurred in the
aftermath of two very important changes in the U.S. communications industry:
divestiture and deregulation. The emphasis here will be somewhat historical in
nature, for two reasons. First, how companies have responded in an organizational
sense to changes in the industry is very much related to the history and evolution
of public and private networks over the last ten years or so. Second, these patterns

are repetitive and built-in to the dynamics of the communications industry; hence, they can be expected to surface periodically as even more changes are introduced into the technological infrastructure and regulatory climate.

In terms of how companies dedicate their own internal staff and resources to using communications strategically, many of the changes leading to the current profile began with divestiture in the early 1980s. In the relatively tranquil world of predivesititure telecom managers, the name of the game was voice communications. Their primary responsibility revolved around maintaining their company's voice communications networks—a job often perceived as a caretaker function. Prior to divestiture, a large part of this job involved routine interfacing with AT&T and administrative functions: overseeing such matters as moves and changes, service provisioning, billing and accounting, and cost containment. With respect to this last item, network managers had perhaps their greatest visibility in terms of senior management. Furthermore, before the nation's telephone system was dismantled, the network manager could count on fairly predictable performance from the world's best-engineered public network. AT&T knew where its bread was buttered and kept its corporate customers reasonably happy, despite the occasional unanticipated outages and glitches.

In this view of the telecom world, voice communications was typically viewed as simply another facet of the corporate infrastructure, necessary but placed in the same category as maintenance of the car and truck fleet or the stable of office equipment such as photocopiers, fax machines, and printers. Organizationally, many telecom managers reported to administrative departments within their companies. It was, for example, not at all atypical for a telecom manager to report to a director or vice president of administration, or some other department that subsumed such functions as personnel, operations, and the like.

The individuals who typically filled these positions were not necessarily well-trained in telecommunications technologies. AT&T took care of the technical side of things, and telecom manager didn't need to understand the intricate workings of the Bell system any more than a fleet manager necessarily has to know how to fix a carburetor. Many telecom managers came into the position not by virtue of technical savvy, but rather as a result of a transition from a similar administrative position, one not necessarily associated with telecommunications. The skill sets required would likely be the same: experience in dealing with an outside vendor, the ability to manage an internal system of resources, an aptitude for cost control and billing issues, and basic "people skills."

In the early 1980s, two events took place that had a major effect on this picture. The first was the beginning of a steep and rising trend toward the use of data communications. The second was divestiture, the dismantling of AT&T's telecommunications network monopoly in the United States, coupled with a parallel trend toward the deregulation of the communications marketplace.

In 1980, most corporations carried a mix of voice and data traffic that was 80-20, voice to data. In 1984, the ratio had slipped to 70-30. By 1989, voice was at 55% and data had spiked to 45%. Beginning in the early 1990s, companies began crossing over the 50-50 threshold, a trend that will surely continue [1]. In yet another perspective, figures from the Yankee Group indicated that, in 1991, voice traffic in the corporation was growing at approximately 6% per employee per year. Data traffic, on the other hand, was enjoying a growth rate of approximately 35–40% per employee per year. This massive increase in the amount of data flowing over the corporate network—much of it mission-critical—served to place new and significant demands on the role of the telecommunications manager. In effect, it meant that in some cases, the telecom manager's job responsibilities were gradually being rewritten to reflect those of a data communications manager.

Another major factor affecting the telecom manager's corporate role was the divestiture of AT&T, under the terms of the Modified Final Judgment (MFJ). Under the MFJ, AT&T's local exchange business was divided into what we now know as the seven "not-so-Baby" Bells, the Regional Bell Operating Companies (RBOCs). AT&T's long-distance business, along with other aspects of the company's operations, of course, remained intact. The pros, cons, and complicated extenuating circumstances of divestiture have been covered well and extensively in other texts, so I propose to avoid replicating them here. References are appended for readers who would like more background on the whys and wherefores involved [2]. Suffice it to say that it was this event—in tandem with deregulation and the rise of mission-critical data communications utilization—that placed singular new demands on telecom managers, moving them out of the relative anonymity associated with a background administrative role and into the spotlight of strategic planning and operation. In many cases, however, the set of managerial capabilities needed to address these new challenges could not always be afforded by tapping either the ranks of traditional telecom managers or the new breed of computer science managers and technicians that were filling the departmental ranks.

A GAP IN THE KNOWLEDGE BASE

What options were available to telecom managers who suddenly needed to learn more about the field of data communications? Unfortunately, in the early 1980s, little in the way of educational resources was available. At that time, there was a small number of books, but very few university programs and alternative educational vehicles that dealt with the subject. It was a new and rapidly growing field but had little in the way of precedent to support it. Furthermore, many universities had not yet developed formalized programs either within or outside of their normal degree programs. At first, one of the best ways to become proficient in the practical aspects of this subject area was via data communications companies that had developed their own training programs. In many cases, these programs were developed on an ad hoc basis to respond to rapidly evolving requirements for staff training.

This educational barrier—or, more accurately, information time lag—is, of course, no longer a problem. As evidenced by the reams of direct-mail marketing literature currently being distributed by major universities and independent educational institutes, there is now clear recognition of the enormous value of data communications. New York University, Boston University, the University of Colorado, Northeastern University, and George Washington University are just a few of the schools that have proactively seized this educational opportunity. New York University, for example, offers a wide range of courses via their continuing education department, with titles such as "Data Communications for Secretaries." Boston University has a joint-studies program underway (in conjunction with Wang's corporate education center) that offers courses in both voice and data communications. And the list goes on.

The real experts in the newly emerging field of data communications were the proliferating numbers of enterpreneurially driven data communications companies, many of which had entered the market to sell devices that could jury-rig the public network for something it was never designed to accomodate: data. Companies such as Codex Corporation (now Motorola Codex), Racal-Milgo, General DataComm, and Paradyne (now AT&T Paradyne) experienced significant growth in the years following divestiture largely by designing and developing modems, or devices that convert analog signals into digital signals, thereby enabling data traffic to flow over the voice-oriented public network. Such companies thrived on building equipment that provided *interfaces* between the existing but obsolete analog network and the rapidly growing base of digitally oriented computer systems in corporations. More so than the universities or even many large telecom providers, these companies—with their experience base in making the public network do "data tricks"—became highly knowledgeable about this rapidly growing field of data communications.

As stated, the public network was never designed to accommodate the kind of data traffic that was now being generated in corporate environments. Accordingly, telecommunications managers suddenly found themselves having to deal with the engineering implications involved in using a network designed for one set of purposes yet needing to be adapted for another, completely different set of functions. Furthermore, those who have working knowledge of both areas will readily admit that the worlds of voice and data communications are about as similar as anthropology and quantum physics. Thus, as a result of this knowledge gap and the widely divergent requirements involved in both voice and data communications transmission, network managers were faced with the task of designing their own corporate networks to accommodate ever-increasing amounts of data, but often without the background, skills, and technical background necessary to do so.

Thus, the onus of designing and developing these new data networks was no longer falling on the public service providers, but on users themselves, even if the transition was a gradual one. When AT&T "went away," telecom managers were no longer able to fall back on a single-source supplier for all their communications needs: local, long-distance, and international. Under the MFJ, AT&T could now no longer provide this kind of "one-stop shopping," and many of the various circuit segments involved in the process of network provisioning previously provided by AT&T were now available only on an unbundled basis.

In many cases, each portion of the network now had to be established separately, and an increasingly bewildering array of both circuits and equipment had to be dealt with. Local exchange circuits, for example, now fell under RBOC jurisdiction and had to be leased or procured separately from the long-distance portions of the network, the domain of AT&T, MCI, and Sprint. In addition, divestiture created a series of local access and transport areas (LATAs), which, in essence, constituted the territories of the twenty-two individual Bell operating companies (BOCs). In each BOC, there was a defined number of LATAs, and the BOCs were required to provide service within each, under the rubric of intra-LATA communications. They were not allowed to provide communications between LATAs, which was defined as inter-LATA or long-distance communications. This piecemeal and somewhat arbitrary division of labor within the public network created a new series of complications for telecommunications managers trying to develop integrated network services within their companies.

THE GROWTH OF PRIVATE NETWORKS

From a practical, day-to-day management standpoint, divestiture and deregulation created some daunting challenges for network managers. In many cases, the infrastructure for new data and voice networks now had to be built piece by piece, whereas previously the AT&T network seamlessly and transparently provided those services. Now the seams were rent, the switching fabric torn asunder, and the public network was becoming a complicated tangle of options and alternatives. The giant "black box" that was the public network became a Pandora's box filled with lost connections, incompatible protocols, and unhappy users who often developed a pragmatic nostalgia for the high levels of network performance they had long been accustomed to. Fortunately, the condition was transitional in nature, though no less difficult for those that endured it from an operational standpoint.

In spite of these difficulties, many corporations saw an opportunity in the making and were pleased with their newfound ability to exercise more control over the design and implementation of their own private networks. These companies began the process of building autonomous networks Chinese-menu style—by choosing circuits from column A, access devices from column B, and modems and multiplexers from column C. Whereas the downside to divestiture was the demise of "one-stop shopping" and the full spectrum of services once available via AT&T, the upside was the freedom to custom-design networks that were suited to the basic strategic and operational objectives of the company, something not possible under the straitjacket of Bell-system uniformity. However, a certain amount of bridge-burning was also necessary. Companies that chose this route now found themselves with the responsibility for designing, developing, and operating these private networks resting entirely on their own shoulders.

The companies that rose to this challenge had thus decided to follow the exhortation to "Be Your Own Bell" (an approach wryly dubbed BYOB). Electing to be your own telephone company implied a lot of pioneering hard work. It also entailed increasing reliance on a stable of competing vendors for the purchase of network equipment that was often functionally incompatible (one of the important luxuries sacrificed under divestiture, of course). In the process of building and operating these new networks, problems would occasionally arise. When they did, network managers often found their vendors in the finger-pointing mode. If there was a line outage or a terminal was down, a modem vendor could walk away from the problem, claiming it was the carrier's fault. The carrier might simply declare that it must be a CPE problem—traceable perhaps to one of the modems configured in the network. And, of course, the same sort of phenomenon could be observed taking place with respect to the RBOCs and various long-distance circuit providers.

This particular tension between public and private networking, taking shape as it did in the aftermath of divestiture, became a major dynamic in the future development of both public and private networks and the telecommunications industry in general. Both AT&T and the RBOCs regretted the erosion of their control of the network as a result of divestiture, relegating them, as it did, to the mundane role of bandwidth provider (that is, providers of the basic transmission capacity that corporations bought wholesale in the building of their own sophisticated private networks). This was especially true of high-end corporate users—the Fortune 1000—who constituted a principal source of business revenue for telephone companies, especially as the use of data communications continued to rise. For the RBOCs and long-distance providers, however, the business of wholesaling bandwidth was not where the action was going to be in the next decade. The future of telecommunications was going to belong to those entities that could provide a network with built-in computer-based intelligence, providing a wide array of interactive services for users. It was this territory that both the public network providers and the major computer makers would find themselves competing for in the 1990s.

WHO CONTROLS THE NETWORK?

Understanding the roles of public and private networks is a precondition to appreciating the organizational dynamics of how companies develop staff, equipment, and resources pertaining to the strategic use of computers and communications. It also provides a perspective on some of the real-world constraints involved in the use of telecommunications as a so-called "strategic weapon." To fully appreciate the public-private network dynamic, it's necessary to peek behind the curtains and view the larger forces at work in the process, namely AT&T and IBM.

For AT&T, divestiture held out the promise of significant new business opportunities in computers and computer communications [3]. Under the new freedoms granted by divestiture, AT&T made the commitment to become a major player in the computing arena. Likewise, the RBOCs realized that the future of networking was predicated upon making the PSTN more intelligent, given that the functionality of computers would increasingly be intertwined with the networks on which they resided. Both AT&T and the RBOCs knew they needed to get out of the pipeline business. The nature of that business was simple enough: selling lines to large corporate users, who then, with their own considerable corporate resources, overlayed those lines with new levels of intelligence and functionality, creating private networks of varying sizes, applications, and complexity. This was reinforced by the realization that unless they moved to create new levels of network intelligence and control for their corporate customers, AT&T and the RBOCs would lose serious competitive advantage as those users moved ahead with their private networking efforts.

Thus, divestiture gave AT&T and the RBOCs access to the emerging new world of computing. This was the implicit *quid pro quo* behind divestiture: in exchange for the breakup of the communications monopoly, AT&T was now free to explore and exploit exciting new market opportunities in the computer arena. AT&T's path to success in this market has been a rocky one, marred by severe financial difficulties resulting from its foray into the PC market but subsequently buoyed by the acquisition of NCR.

In the meantime, IBM was another company keeping a wary eye on the market forces, dynamics, and trends epitomized by the word convergence—the melding of computer and communications capabilities. The so-called battle between IBM and AT&T is, of course, an oversimplification of a basic industry dynamic; many other companies were involved as well, on both sides of the fence. Nevertheless, the idea of a battle tends to serve as a useful metaphor for describing the interplay between the converging forces shaping the direction of the marketplace. Taken in its extreme, there are two diametrically opposing views behind it: "IBM's view of the network is that of intelligent devices communicating over a dumb network. AT&T's view of the network is that of dumb devices communicating over an intelligent network" [4]. However, as eloquently argued in Peter Huber's study of the public network, *The Geodesic Network*, neither of these dichotomous views provides an accurate perspective on the likely evolution of the public network. Instead, the Huber paradigm suggests heterogenous, complex, but dynamically unified intelligent networks over which a wide variety of devices will communicate with great ease of use and transparency, representing elements of both views [5].

Behind the endless jockeying for position and strategic maneuvering involved in this new network paradigm is a key consideration: whether network control should reside with public network providers or is best relegated to the end user. Equipment providers that focused on offering customer premises equipment (CPE) also attempted to provide their customers with ever-increasing levels of network control. This was, of course, essential in the building of private networks, the performance of which needed to be monitored and maintained on a constant basis. In building private networks, the deployment of devices such as T1 multiplexers and packet switches that allowed independent voice or data streams to be integrated into a single line offered several advantages. The first was allowing the user to enjoy significant cost savings by optimizing the use of a transmission line, because routing more traffic over a single line reduces dependency on multiple lines (and their associated costs). Second, better network management of the consolidated traffic stream was now possible. It's clear that the business interests of the multiplexer provider were set in opposition to those of the carrier, who wanted to provide multiple lines to the customer. For the corporate user, of course, the ultimate criterion in this "make-or-buy" decision is based on the long-term economics after equipment payback and other considerations are factored in.

Another major area of conflict and competition between equipment providers and carriers centers on network management itself, a logical point of convergence. Network management comprises a diverse array of operations and applications and, indeed, means something completely different to every individual. As a result, the subject of network management continues to be covered with layers of confusion. (Chapter 6 discusses network management in considerably more detail.) For some, network management signifies the test and measurement of communications signals and parameters. For others, it has to do with the management of staff, resources, and equipment. Yet another meaning relates to performance and reliability and the process of having communications systems available for automatic backup and redundancy when primary systems fail.

For vendors and carriers, however, network management often means something else entirely: strategic account control. Thus, in the battle of private and public networks, network management has traditionally been a major linchpin. The principle here is simple: whoever controls the network controls the account. Because network management systems (NMSs) generally represent a common user interface dovetailing with all other systems and equipment in a customer's network, their strategic importance has become well-recognized by vendors and carriers. In an attempt to seize the high ground, IBM unveiled a network management system called NetView. AT&T counter-punched with its own system, Unified Network Management Architecture (UNMA). Not to be outdone, Digital, the world's second-largest computer maker, developed what it called its Enterprise Management Architecture (EMA). However, as will be discussed later, even these systems will be challenged by a flock of new LAN-based systems now coming onto the market.

Since those announcements several years ago, a number of major carriers have also responded with their own proprietary systems, such as MCI with FocusNet and British Telecom with its CONCERT system. This represents an attempt by the carriers to fight back against equipment-based systems residing on the customer's premises. If, for example, carriers can manage to provide users with sophisticated levels of network control, then they can deliver the best of both worlds: the economies of scale and attendant cost savings afforded by leveraging the installed base of public network systems through virtual private networks, and the ability to manage them with the same agility that users came to appreciate and eventually require in building their own private networks.

In the late 1980s, the "battle" between public and private networks began to shift into a new phase, characterized by more aggressive approaches to marketing and sales on the part of the carriers, no longer content to sit back and be pipeline providers. Much of this new aggressiveness could be traced to increasingly heavy competition among long-distance carriers, especially the "big three:" AT&T, MCI, and Sprint. The target of this competition was the most important revenue-generator for long-haul providers: the large corporate user. An important part of the marketing strategies employed by all three had to do with the carrier's ability to let customers control and reconfigure their particular constellation of network services from their own premises.

The concept of allowing a user to directly control service provisioning and performance via an on-premises terminal was originally called customer-controlled reconfiguration (CCR), as pioneered by AT&T. CCR effected a major change in the carrier-customer relationship: instead of relying on account representatives to make every little change in the design and provisioning of a network service, customers were provided with terminals on their own premises. Via this device, the user could make network service changes in real time, a process that offered far more flexibility and independence from the carrier. Lines could be added or deleted to network service, or paths reconfigured in the case of virtual private network (VPN) arrangements. In addition to control and reconfiguration, such systems could also provide line status and other performance-related information. In this way, the customer's network technicians could view the same line performance statistics seen by technicans stationed in the central offices of the carrier. RBOCs, such as BellSouth, have also been experimenting with such approaches.

These aggressive marketing moves by carriers, as they play catch-up with private network capabilities, are slowly moving the telecommunications industry closer and closer to the ideal of the all-digital, switched, universally available high-

PSTN= Public Switch Telephone Network

bandwidth network that was the underlying concept behind the ISDN. In AT&T's parlance, this vision of future public network capabilities was called universal information services (UIS). Whatever term is used, it is clear that the public network is, in fact, evolving to greater levels of intelligence, flexibility, ubiquity, and performance. Eventually, the PSTN will, at a minimum, possess the following characteristics:

- High bandwidth capacity;
- Fully digital transmission;
- Increased applications intelligence;
- Greater bandwidth-on-demand flexibility via switched services;
- Migration toward SONET-based fiber-optic transmission;
- Migration toward integrated (voice, data, video) broadband switching;
- Continued provision of user-controlled network management.

THE OUTSOURCING QUESTION

As the rather tortuous but steady building of this new model of the public network continues, users will, as in the past, avail themselves of public network services when doing so makes good economic and managerial sense. However, the tug-of-war that took place between public and private networks is not over. The cycle will continue to repeat itself as new technologies such as SMDS, frame relay, SONET, and ATM-based broadband ISDN are deployed in both private and public networks, with capabilities in some cases first appearing in the former. However, as the tortoiselike carriers begin to catch up with the harelike and more market-agile equipment providers by implementing these same capabilities into the network, the balance of power may indeed shift over to the carriers, if accompanied by a supporting shift in market economics. Such important regulatory factors as rate of return versus price cap regulation for both local-exchange and long-distance providers are major variables in this complex equation, as are issues such as Tariffs 12 and 15 with respect to AT&T.

In the meantime, to what degree users should maintain control of their networking resources, as embodied partially in the public versus private network debate, will continue to be a valid issue. *What's at stake for users is a basic question: Is the corporate telecommunications network too important a strategic asset to outsource to external providers?* In Chapter 6, outsourcing is explored in terms of a classic "make-or-buy" decision that faces companies. However, outsourcing, like network management, comes in many shapes and sizes. One form of outsourcing does indeed involve turning much of a user's control of the existing network over to a new breed of global service providers, who are aggressively going after the business of multinational corporations. Examples include Sprint, British Telecom, France Telecom, AT&T, NTT, MCI, and Cable and Wireless.

Companies faced with the prospect of outsourcing have a wide array of choices. They can outsource either computer operations or telecommunications, or both. Within those individual categories, such functions as planning and design, network management, applications development, user support, equipment maintenance, and ongoing operational oversight can all be outsourced. Companies exploring the possibility of outsourcing will also be faced with the prospect of what kind of major outsourcing provider should be used. Among the choices are not only the globally competitive major carriers cited above, but also the large computer makers and traditional systems integrators such as EDS and Arthur Andersen. These choices are complex and involve important and fundamental strategic decisions about whether to hand over many important network functions to an outside entity that may not fully understand the company's business charter, long-range objectives, and specific networking needs.

NOTES AND REFERENCES

[1] T. Zerbiec, "Private Network Challenges and Opportunities for the 1990s," *Telecommunications*, January 1991.
[2] For more on divestiture, see the following:
   Robert W. Crandall and Kenneth Flamm (eds.), *Changing the Rules: Technological Change, International Competition, and Regulation in Communications*, The Brookings Institution, Washington, D.C., 1989.
   Alan Stone, *Wrong Number: The Breakup of AT&T*, Basic Books, New York, 1989.
[3] Thomas S. Valovic, "Public and Private Networks: Who Will Manage And Control Them?" *Telecommunications*, March 1989.
[4] Private communication with Ken Bosomworth, President, International Resource Development (New Canaan, CT).
[5] Peter Huber, *The Geodesic Network: 1987 Report on Competition in the Telephone Industry*, U.S. Department of Justice, Washington, D.C., January 1987.

Chapter 6

IS/Telecom Organizations: Managing the Unmanageable

Amount of money transferred electronically every minute among U.S. banks: $819,300,000.

—*Harper's Index*

U.S. corporations are increasingly learning how the strategic use of telecommunications can add value to their organizations. Many important and useful new applications are being developed by forward-thinking companies that have recognized the potential inherent in this new set of business and marketing tools. During the 1980s, however, the early adopters—the proverbial pioneers with the arrows in their back—had learned some difficult lessons. Both senior and middle management were on a steep learning curve regarding the complexities of the new technology and what kinds of business applications it could, couldn't, and shouldn't support. There were also issues of corporate culture at stake, along with the resistance to change that accompanies any radically different approach to traditional modes of conducting business.

As has become increasingly evident, the results of these early IS (short for information services, formerly and sometimes still known as MIS) implementation efforts often amounted to little more than automating routine business practices and procedures. As a result of the mismatch between ideal and reality, IS departments in many cases lost points with senior management or gained a reputation for failing to appreciate the real business impact of the systems they were implementing.

The problems associated with how computer and communications technologies have been implemented in the business environment persist to this day. They are directly related to the fact that computer professionals, by nature of their training and professional practice, tend to focus on the operation and maintenance of computers. They are not marketing managers, accountants, engineers, or personnel directors. Nor has their training in computer science prepared them to understand the wide-ranging and rapidly changing requirements of various corporate functions. Such training, of course, would be extremely useful in helping IS professionals better translate their valuable knowledge into useful applications by appreciating the day-to-day needs of corporate departments that are, in essence, their customers.

As a result of this confluence of events, the leap of faith that many senior managers made in response to the much-hyped promise of emerging computer capabilities often seemed, in retrospect, more like a leap into a dark technological abyss filled with arcane buzzwords, horrendous cost overruns, poorly functioning systems, unfulfilled target goals, and unhappy users. In fairness to IS, many expectations concerning the role of computers in the corporation had been raised to unrealistically high levels. There were many factors that contributed to this, including but not limited to the aggressive promotional activities of the major computer-makers and other players trying to keep up with them; the magnification and distortion of marketing overreach fostered by consultants and some of the trade press; and generally excessive levels of industry technobabble, hype, and jargon. As a result, once reality surfaced in the form of performance delivered for dollars spent, the corporate view of IS was subject to a serious adjustment and never quite regained the aura of technological mystery that sustained it during the initial period. Here's how one observer described it: "As the heady technology days of the late 1980s fade into the sober 1990s, the honeymoon glow surrounding corporate executives and information technology has sharpened into the harsh glare of everyday life. "Gee-whiz" attitudes and the blind pursuit of competitive advantage have matured into clear-eyed realism. The new focus is targeted, functional, and cost-justified" [1].

Thus, the view of IS was tempered by this new realism on the part of senior management. Much had been made of the need to invest in a state-of-the-art information service and communications infrastructure. Many companies found themselves unrealistically caught up in the fever to use IS for competitive advantage without having either a strategic corporate vision or a tactical and pragmatic operating plan that focused on getting back a reasonable return on investment. As a result of these developments, the early 1990s brought many companies into a period of reassessment and the attempt to see, in a more objective fashion, exactly what IS/telecom was delivering to the corporate bottom line.

The results were sobering. As a result, many companies set out to make changes in their organizational structure and to explore techniques such as business reengineering to restructure not only their IS systems, but more importantly and more fundamentally, the business procedures that they supported. Other changes pertained to the organization itself. Some companies, for example, began to reassess the kind of professional they had thought was needed to head up an IS department and the basic skill mix that such an individual might optimally possess. In some cases, the judgment was made that this mix was skewed toward an overabundance of technical savvy and not enough business background. In an effort to correct the problem, a discernible trend took hold in favor of hiring nontechnical executives for the position of chief information officer (CIO) or its equivalent: "While the ideal CIO would have strong management, business, and technology skills, it is often difficult to find all those qualities in a single executive. As a result, some companies are handing the CIO job to managers with more business than technology savvy because of concern about the perceived inability . . . to translate computing and network assets into a corporate advantage. Others have opted for a two-tiered hierarchy in which the business-oriented CIO is supported by an experienced IS manager" [2]. The impetus to bring a more business-oriented individual into this role was driven by two major objectives that became very important to IS departments in the 1990s: controlling skyrocketing costs and deploying IS systems for better competitive advantage.

THE ROLE OF TELECOMMUNICATIONS

Prior to divestiture, telecommunications management was a fairly routine function in most corporations. It had none of the aura or mystique about it that IS activities enjoyed and often fell into the same operational category as other in-house services such as fleet management, mail service, building maintenance, and so on. In many cases, telecommunications department heads reported to a director of administration or a similar position. As telecommunications became increasingly important to the corporate organization during the 1980s, it was sometimes subsumed under the organizational umbrella of the IS department. In other cases, it continued to remain in the realm of administrative services or, as frequently happened in financial services companies, it developed its own hierarchical channel terminating in a senior management position such as vice president of telecommunications.

In 1984, AT&T was divested into seven regional companies under the conditions of the Modified Final Judgement (MFJ). This event shook the telecommunications industry to its roots and liberated telecommunications professionals from their relatively mundane roles, propelling telecommunications issues to the top of the corporate boardroom agenda. In this environment, the world of five-year operating plans, corporate strategies, and senior executive attention and interest opened up many new opportunities for telecommunications professionals as well as presenting a series of daunting challenges.

The role of the telecommunications manager has evolved in a rather dramatic fashion since divestiture. As more and more core business activities came to be transacted on large corporate networks—especially mission-critical applications such as those found in banking and insurance—telecommunications management grew proportionately in terms of level of responsibility and organizational size. Throughout the late 1980s, major world financial centers such as New York City developed an increasing dependence on the use of telecommunications for such operations as funds transfer between accounts at domestic and international locations. Coupled with this was an increasing dependence on time-sensitive electronic information for the purposes of 24-hour trading in world financial markets. In the financial community, the rapid rise of automated teller machines (ATMs) was a major factor in the proliferation of relatively low-speed data communications networks. Such networks gradually extended their tentacles throughout major urban and suburban areas, bringing the bank to the customer instead of the customer to the bank. Interbanking systems, such as Cirrus and NYCE, were developed and implemented, allowing individuals traveling in various parts of the country to access funds electronically from banks at their home location.

During this time of rapid network growth, the information-intensive financial and services industries tended to be the leading-edge users of telecommunications services. From one point of view, financial transactions were a logical starting point for IS applications because, in their purest form, they represented a relatively simple exchange of one particular type of information. Such information was ideally suited for adaptation to an electronic medium that could reflect the time value of money far more accurately than had ever been possible. With this increase in financially oriented electronic data also came what futurist John Naisbitt has called the "collapse of the information float," whereby the "float" or transit time involved in the movement of paper assets from one account to the next could be shortened and consequently exploited in the pure economic sense as additional revenue. Thus, the first aspect of this transition to the long-predicted cashless society was to make transactions electronic. Increasingly, however, more of the actual instruments of traditional business transaction will also become electronic. For example, the use of imaging technology in the banking industry is well underway, eliminating the need for checks to be physically transferred from one bank to another.

During the 1980s, increasing amounts of sensitive and mission-critical types of data traffic found their way onto corporate networks, making telecommunications an absolutely indispensable part of a company's day-to-day operations. That being the case, the principal operational devil in the telecom manager's professional world could be summed up in one word: downtime—that is, the amount of time that a network or some essential portion of it is taken out of service. Minimizing the impact of service disruptions and systems downtime thus became one of the principle measures of a telecommunications manager's on-the-job effectiveness.

For major financial institutions such as Chase Manhattan, Chemical Bank, or Citicorp, downtime came to be measured and quantified in very specific terms. For example, a typical measure of downtime might be on the order of $100,000 per minute (that is to say, for every minute the network was out of service, the company lost approximately that much in revenue.) This loss might come in the form of orders that couldn't be transacted and signed/sealed due to the dependence upon linked computers, or other more complex transactions involving the time value of money. In this context, the old piece of conventional business wisdom that "time is money" began to take on an entirely new and distinct meaning.

When mission-critical networks are involved, telecom and network managers tend to be preoccupied with two important aspects of network performance: *reliability* and *availability*. Reliability is an engineering term frequently used in the aerospace industry to connote the overall performance effectiveness of a given technical system or subsystem. It is frequently expressed as a percent of optimal performance, as in the case of NASA's often-cited 99.9% quality assurance levels. A typical measure of the reliability of a system or piece of equipment is mean time between failures (MTBF) or mean time to failure (MTTF). Availability—the converse of downtime—is a close cousin of reliability and is generally used in reference to user applications rather than specific systems or equipment. Availability is what the in-house user looks for in the network: a system that's up and running at optimal efficiency whenever it's needed. In this sense, availability is to the user what reliability is to the network manager.

These performance-oriented goals became important enough to justify the establishment of telecommunications departments of significant size as management worked to ensure that downtime didn't hurt corporate revenues. Financial-services giant Merrill Lynch, for example, developed a telecommunications staff level of approximately 150 to cope with their increasing involvement in the world of telecommunications, as both an in-house and external provider of telecom services through their former Teleport subsidiary. Typical sizes of telecommunications departments in other financial organizations ranged from 25 staffers to as many as 200 within organizations such as the major New York–based banking conglomerates.

Given these developments, it's clear that in certain organizations in which telecommunications was indeed mission-critical, the telecom manager's job had moved light years away from the days of making sure that the internal voice phone network wasn't malfunctioning, and it was the advent of that increasingly precious business commodity—data—that precipitated the shift. In larger telecommunications organizations, the telecommunications function often achieved executive-level status. Citicorp, for example, has both a vice president of telecommunications operations and a vice president of regulatory affairs for telecommunications. The status of these positions reflects the overall importance of telecommunications to these companies and their core business activities, underscored also by the large number of professional staff involved. However, as telecommunications became more organizationally significant and was thrust into the spotlight of senior management attention, certain dimensions of the telecom manager's job remained distinctly similar to the more mundane aspects of the predivestiture role. Essentially, the nature of the management function was to keep things running smoothly, with the focus of corporate attention intensifying only when problems developed. As a result, in many companies, the major incentives for performance continued to be keeping costs down and keeping the network quietly and invisibly operating— both risk-aversive rather than risk-encouraging activities, which shared aspects of the caretakerlike expectations of an earlier time when telecommunications had a lower profile. Whereas the strategic use of telecommunications was indeed growing in importance throughout the 1980s, telecommunications managers often found themselves called upon to establish the infrastructure for corporate networks and make sure that they operated efficiently and cost-effectively, rather than being asked to spend the majority of their time devising innovative strategic applications.

THE STRUCTURE AND FUNCTION OF TELECOM DEPARTMENTS

Throughout a spectrum of vertical industries, there is considerable variance in how telecommunications departments are structured, staffed, and positioned in the corporate organizational hierarchy. Although it's difficult to generalize, there are nevertheless some distinct patterns that emerge. For example, a typical telecommunications staff in a Fortune 1000 insurance company might have a staff of approximately 25 to 100 telecommunications professionals, all of whom report to a director of telecommunications. That director, in turn, might report to a VP of telecommunications or, if none existed, a VP charged with administrative responsibilities. Of this group, approximately half might be involved in what could be described as network operations, or the actual monitoring and control of network functions. This includes manning help desks, the phone-in centers designed to give individual users immediate response to network problems and outages, often on a 24-hour, seven-day basis; and in general, keeping the networks up and running. This group would also be the one most involved in interfacing with the end user (or internal customer). Another 15% or so might be involved in network engineering functions, including recommending and specifying equipment, and the task of ensuring that various portions of subnetworks were operationally compatible. Another 15% would be involved in network planning and expansion activities, and the remainder would be dedicated to administrative duties such as keeping track of inventory, cost and accounting functions, and so on [3].

What about the network manager's position? Is there any way of characterizing or typifying its profile, given that throughout that same spectrum of vertical industries, that profile will tend to vary significantly? One survey dealing with this question attempted to determine some of these characteristics [4]. It found that a typical network manager had ten years of experience in telecommunications, a bachelor's degree, earned approximately $58,000, and held his or her current position for slightly more than two years. The survey also showed that responsibilities for voice and data among these managers tended to be mixed. Approximately 55% were identified as being responsible for data networks only; 37% indicated responsibilities for both voice and data. Only 8% of the respondents had responsibility for voice communications only. Finally, more than 80% of those responding to the survey said that they also had responsibility for LANs as a part of their normal job responsibilities.

Another area worth exploring is the survey's breakdown of educational levels. This information shows that the variance in backgrounds is considerable. This is traceable to the fact that the career path of network managers is by no means a fixed route, but rather tends to draw on a widely disparate talent pool. As many network managers tended to have technical degrees in computer science as degrees in business administration (18.4% for the former; 16.5% for the latter). Even more significant is the fact that certain types of educational backgrounds such as mathematics (5.6%) and accounting and finance (4.7%) weigh in far more heavily than degrees in the field of telecommunications itself (2.6%). As discussed earlier, this is a reflection of a time lag and requirements dissonance that still exists between the educational process and corporate staffing requirements.

Another important source of information on the professional characteristics of telecommunications managers comes from a survey conducted by the Wall Street Journal [5]. That newspaper's survey of telecommunications decisionmakers belonging to the International Communications Association (ICA) provides some additional insights into the nature and scope of their responsibilities. According to the survey, 100% of the respondents indicated that they were at some level involved in corporate decisions to lease or purchase telecommunications equipment and services. Concerning the involvement of MIS and other senior corporate management functions in the decision-making process, the survey had this to say:

> Top management is most likely to become involved in telecommunications leases or purchases according to 98% of those surveyed. Among top management with telecommunications responsibilities, telecommunications directors and MIS/DP directors were selected by the largest number of decisionmakers as most likely to be involved in purchase decisions. In addition, vice presidents and financial officers were reported to be involved in decisionmaking by four out of ten respondents, indicating that decision-making involves a wide range of corporate top managers. [6]

Concerning the nature of the telecom manager's job itself, the survey revealed some other interesting dimensions. First, 11% of those surveyed held the title of vice president of telecommunications. The majority of those responding (43%) fell into the category of telecommunications manager, with another substantial group (27%) indicating director-level responsibilities. (Note also that the survey points out that the term "telecommunications" has been used to include any titles with the words data, voice, or network in them.) Finally, the survey provided information on the types of vertical industries that these telecommunications managers are most likely to be involved with. The largest segment here was manufacturing (43%), followed by financial insurance and real estate (close to 20%) and services (14%).

In the late 1980s, a new trend began to emerge among IS and telecommunications departments throughout a broad spectrum of industries: outsourcing. As discussed in Chapter 5, outsourcing is a management strategy option whereby portions of the management of IS and telecom functions are turned over to third-party service providers, systems integrators, large computer makers, or some other outsourcing entity. Under this approach, the outsourcing provider might assume responsibility for specific, contractually determined aspects of the design, implementation, or ongoing day-to-day management of computer operations and telecommunications functions.

IS/TELECOM AT THE EXECUTIVE LEVEL

One of the most important aspects of telecommunications management is its relationship to the IS department and related activities. Just as the telecommunications function has risen in stature and corporate involvement over the last ten years, so has the role of the IS group. In fact, IS managers were able to penetrate the upper echelons of senior management (VP levels) much earlier than did their telecom counterparts. Out of senior management's recognition of the IS manager's newly emerging role came the concept of the CIO, or chief information officer. Much has been written on the CIO concept, both positive and negative; accordingly, it will not be covered here in any great level of detail. However, as with many other phenomena in the corporate arena, it's worth noting that perception doesn't always match reality.

The CIO concept elevated IS management to the level of vice president, with attendant access to chief operating and executive officers. This shift in organizational perception afforded an opportunity for IS managers to assume a far more dynamic role in the shaping of corporate policy. However, with the benefit of hindsight, the CIO concept now seems more a gleam in the IS manager's eye than the basis for championing a new organizational *wunderkind*, who, with support and encouragement from senior officers, would transform the very nature of the corporation from the inside out. In principle, the basic idea was that a CIO would have planning-level access to the corporation's inner circle—essential if computers and communications were ever to assume more significance as strategic tools for reinventing basic business operations. In practice, however, there were unanticipated glitches. For one thing, CEO-CIO relationships often proved to be complicated, difficult, and burdensome, as each function struggled to understand the requirements, demands, and challenges of the other. To a certain extent, the technology itself—still the untranslatable and intransigent unknown to many CEOs—was the source of this difficulty. This was compounded by the fact that many CIOs didn't have the requisite business savvy. Both of these factors naturally paved the way for a classic breakdown in communications. Another factor, of course, was that many CEOs began to suspect they were getting little return on investment for the millions of dollars sunk into these sprawling, budgetary black holes called MIS systems. By 1988, it was estimated that only 10% of IS organizations in the United States had individuals installed in the position of CIO, and in the years that followed, this ratio would not increase significantly.

At some point in the process of defining the CIO concept, another new and fairly radical variation on this theme emerged: that of chief networking officer, or CNO. Although some observers considered this apparent one-upmanship over the CIO concept slightly preposterous, others saw it as the wave of the future. There were several justifications for this latter view. First, borrowing from the marketing slogan used by Sun Microsystems that "the network is the computer," the deployment of computer resources was expected to become increasingly network-oriented throughout the 1990s. Such an outlook, in fact, closely approximates the technical reality because highly internetworked computer resources do act as a single entity, especially from the standpoint of the end user. Furthermore, whereas in the 1980s the principal challenge for corporations was to deploy computing resources, the challenge in the 1990s was to link those resources—many of them duplicate—to create optimal efficiencies of both cost and performance. Second, the term CNO garnered favor because it brought telecom into the forefront in a dramatic and convincing fashion. In that sense, it would be overlooking a fundamental reality not to note that in the day-to-day operational trenches where IS and telecom professionals commingled, there was a long-standing breach of cultural understanding between the two groups. This was certainly understandable, because the two camps viewed the world from very different vantage points and yet were increasingly forced by the circumstances of convergence to work together to accomplish the same (or at times overlapping) goals. Just as the IS function strove for legitimization and acceptance in the corporate inner circle via the CIO concept, telecommunications management seized upon the CNO concept as a vehicle that might lead to greater visibility in the corporate organizational structure. The CNO concept—although somewhat less realistic than that of CIO—represented a means for moving closer to this goal. It is unlikely, however (at least for the next five years or so), that the idea of the CNO will become anything more than just an interesting construct.

Given these developments, how did telecom organizations generally relate to the IS departments under whose managerial shadow they often fell? Survey statistics again can shed some light here. For example, in the first survey cited earlier, it was found that approximately 37% of the organizations surveyed had telecom departments reporting administratively to IS functions. However, many organizations (41%) maintained a separate budget from that of IS. Other key reporting relationships included administrative (15.5%) and miscellaneous (20.8%). These data clearly paint the picture of an intracorporate function in a state of flux and lacking a great deal of consistency across the spectrum of vertical market industries.

STAFFING: WHO WATCHES THE NETWORKS?

One of the most persistent problems for telecommunications managers in the late 1980s was staffing. Whereas IS departments could be assured of a steady stream of applicants with freshly minted degrees in computer science, telecommunications had no such equivalent advantage to fall back on. As a result, as networks grew in size and complexity, the number and quality of professionals available to cope with this impressive growth remained relatively constant.

In terms of educational resources, there are a number of universities that have consistently maintained strong telecommunications departments over the years. (The University of Colorado is one prominent example.) For the most part, however, telecommunications as an academic discipline has not been widely adopted by many institutions of higher learning. The Massachusetts Institute of Technology (MIT), for example, does not support a full-fledged telecommunications department, and many other well-known universities have chosen not to develop undergraduate, much less graduate, programs in telecommunications. In those schools at which telecommunications has developed a certain amount of prominence, the emphasis tends to be on policy and regulatory issues, rather than focusing on a technical and managerially oriented curriculum. These kinds of approaches are often outgrowths of long-standing educational programs that tended to emphasize broadcasting and mass media issues (such as radio and TV). Such programs often piggyback on the curriculum of an associated discipline. For example, in MIT's case, the political science department for many years ran a small program that was limited to the study of communications policy. Columbia University and Northwestern University's Annenberg School are examples of other institutions that have traditionally emphasized regulatory policy over the development of technical programs.

Although such programs are important and turn out professionals competent in assessing the impacts associated with public policy issues, the kinds of skills and abilities they impart are unfortunately of marginal value in the context of mainstream telecom management. From the standpoint of a telecommunications manager who might be eager to add capable individuals to existing staff, degrees in telecommunications policy, although more desirable than other disciplines, are often inappropriate (although such backgrounds might indeed be useful in positions related to regulatory affairs).

As the educational availability–corporate demand mismatch continued, competition for telecom staff became increasingly fierce as companies lost personnel to the opposition. Often this occurred as the result of the pressures of competitive bidding and the shrinking of the available labor pool. To make matters worse, as the numbers of available professionals diminished in real terms, the functional advancement of network management systems—a technical fix that had the potential to alleviate staffing pressures by automating certain routine functions—remained on the slow track. In any event, network managers found that they could not get as much help as they had once hoped from network management equipment or services in coping with the increasing problem of staff shortages. In fact, in some cases not only did the available network management equipment provide no measurable relief, but the deployment of increasing numbers of these systems actually aggravated the problem because the result was only more "tubes" to monitor and not enough people to assign to that task.

Those companies that were able to somehow adequately address the issue of staffing requirements often found themselves hiring individuals with a wide (that is, uneven) range of skills and abilities. This became especially pronounced as hiring crossover was at its height, because each vertical industry tends to have its own particular set of application requirements. For example, the retailing industry required particular business applications expertise in the areas of purchasing and inventory control, in addition to the basics of developing various telecommunications options to enhance those functions. But the real value of a professional was at the intersection of expertise in communications and the applications involved. Thus, when a professional moved from retailing into another corporate applications area—for example, insurance—the basic level of telecommunications capabilities were transferrable, but applications-specific knowledge would have to be relearned. This was an entropic situation, worsening in direct proportion to the number of vertical market crossovers.

To make matters worse, much of this applications-specific knowledge could be learned only one way: from experience. Thus, experience within a given industry could disproportionately increase the value of telecom professionals in their chosen area, inflating the market and causing a rash of job-jumping across the spectrum of vertical industries. This particular hands-on aspect of emerging staff capabilities in the telecommunications field represented a distinct delimiter in the quest to develop adequate departmental staff. Thus, even if the universities could have churned out a sufficient number of graduates to meet the requirements of telecommunications departments in the Fortune 1000, the staffing problem would have likely continued to exist. As several observers put it:

> Managing a modern communication organization also calls for the daily
> utilization of various financial skills, since communication costs directly

effect the bottom line of virtually every size company. Since the communication industry is changing so rapidly, the only way to acquire some of the necessary knowledge is through direct, on-the-job experience. The pool of professionals with all of these desired (necessary) qualifications has become smaller and the costs associated with "growing your own" is very high. [7]

Intriguingly, just as the software industry came to face its biggest challenge—the people challenge—so did the burgeoning growth of networks arrive at a similar obstacle. The level of expertise available via the usual sources (that is, the typical churn of the business-education cycle) was no longer sufficient to provide the capabilities necessary to run networks on which mission-critical applications were increasingly being placed. With the two axes of these trends veering in opposite directions, it became clear that traditional telecommunications (and IS) departments would need to begin to draw on a talent pool that existed outside of the normal channels. This is one of the principal drivers for the so-called outsourcing phenomenon—in part, a decision to rely on experts and consultants.

THE PHENOMENON OF OUTSOURCING

As a result of these factors—staffing shortages, the lack of equipment-based network management capability, and the runaway growth of often incompatible networks and computer systems—a major shift in the traditional approach to developing network capability took place in the late 1980s. This shift saw corporate IS and telecommunications departments begin to contract out certain functions and activities that had traditionally been performed by existing in-house staff. The term that came to be associated with this trend was outsourcing. A number of companies—most notably, Merrill Lynch and Eastman Kodak—were early adopters and led the way in experimenting with outsourcing. In many cases, such experiments had interesting results and outcomes, not all of them advantageous in the long run. Eastman Kodak, for example, caused the industry to sit up and take notice when it hired Computerland to manage its PC networks and DEC to manage its *voice* network—a fairly atypical approach. The industry also watched warily when Merrill Lynch outsourced significant parts of its network amid considerable fanfare and then eventually declared the experiment unsuccessful.

The move toward outsourcing also had significance as a risk-aversive action taken by those responsible for operating large computer networks, allowing them to shift the locus of responsibility away from their own operation. The third parties called in to take on these responsibilities represented a wide spectrum: traditional systems integrators such as EDS; management consulting firms such as Arthur Andersen or Booz Allen and Hamilton; or major computer and telecommunications providers such as IBM, DEC, or AT&T. (Some of the latter companies saw major revenue growth from these kinds of activities, which served to significantly offset diminishing revenues from hardware sales.) A major advantage for IS or telecom managers opting for the outsourcing route was that even if the third party couldn't immediately summon the specific expertise required to solve a networking problem, their involvement in the marketplace and resource-intensive professional staff capabilities provided a reasonable level of certainty that the right solution could be made available. This so-called "comfort factor" should not be underemphasized in the complicated process of building mission-critical networks under sometimes adverse and uncertain conditions.

Throughout the early 1990s, outsourcing continued to be the subject of much discussion and scrutiny as IS departments looked for new ways to solve old problems in what were often difficult economic circumstances. A number of the more widely publicized outsourcing moves involved companies that turned over the management of IS or telecommunications functions lock, stock, and barrel to a third party. In some cases, professional staff from the company doing the outsourcing hopped onto the payroll of the outsourcing provider. Yet cases of this type of across-the-board outsourcing were actually quite rare. According to the management consulting firm Arthur D. Little, companies that have opted for a total handoff of staff management functions are often characterized by cash flow problems or are the object of leveraged buy-outs [8]. There is also a trend toward outsourcing at the departmental level, in which a group has experienced continuing dissatisfaction with the responsiveness of their own IS organization. In other cases, certain specific functions within the spectrum of management activities might be turned over to an outside provider.

One interesting aspect embedded in the outsourcing debate relates to the fact that many IS functions are moving toward recentralization, not so much on the operational side, but in terms of development and policy planning. In some cases, this can be related to the fact that, with IS-telecom expenditures often approaching 5–10% of annual revenues, senior management is becoming more involved and putting pressure on IS departments to shape policy more proactively. Another major driver here is the proliferation of renegade LANs and PC-based networks, creating a wide range of disparate and often incompatible networks and a management nightmare for centralized IS functions. Yet, the trend toward departmental outsourcing remains disruptive to recentralization in many organizations, and unless IS departments become more adept at developing higher levels of responsiveness for their "internal customers," the tug-of-war that is taking place over network control may continue for some time to come.

Another trend fueling outsourcing was the need to streamline data center operations. In the great "blank check" expansion of computer systems and technology that took place during the 1980s, companies frequently developed multiple data centers that were deployed across a wide geographical area. Eastman Kodak, for example, had as many as seventeen data centers spread throughout their various

U.S.-based facilities. However, changes in the nature of computing operations, combined with plummeting long-distance costs, created the need for new approaches to the distribution of corporate computer resources. Part of the solution was to combine multiple data centers by consolidating operations. In many cases, outsourcing provided an excellent vehicle for accomplishing this task: a third party could operate more effectively, lacking impediments associated with corporate politics that an internal organization might have to face. Another factor had to do with the high expenses associated with site software licenses, which could be considerably diminished by the process of consolidation.

The decisions that a company faces when considering outsourcing have significant implications for the strategic use of computer and communications resources. Under one theory, in order to use communications strategically, a company needs to keep a tight rein on its own communications capabilities and facilities. Handing over responsibility for the day-to-day planning and operation of the corporate utility network is perceived as antithetical to that goal. However, there are other aspects to this conventional wisdom that need to be considered. For example, it's useful to distinguish between turning over the physical operation of the network and planning and management aspects. In some cases, one of the primary reasons for an outsourcing decision is to allow management to focus more on the latter and less on the operational mine fields involved in running the network on a daily basis—often a sinkhole of time and energy that can keep staff so preoccupied that they'll never be able to shift into a strategic planning mode. There is a strong argument to be made for outsourcing when done in this limited fashion. However, whereas turning over the full spectrum of management tasks to an outside provider may have all the benefits of a short-term fix, it may also prove onerous—and even risky—in the long run. The risks inherent in this approach have much to do with an erosion of in-house staff capability associated with developing strategic applications, which, as has been discussed, is a hard-won item. Moreover, doing so locks a company into a dependency mode with a supplier or systems integrator, which may not be easy to reverse if the decision is made to cut the cord at some later date. All things considered, the jury is still out on the ultimate advantages and disadvantages of outsourcing.

NETWORK MANAGEMENT: THE PROMISE UNFULFILLED

The subject of network management systems and services is an important one in the context of strategic communications. Like outsourcing, network management is a way for managers to direct less time and attention to the physical performance and operation of corporate networks, in order to place renewed emphasis on optimizing their use and developing new applications in support of corporate business objectives. Generally, however, this technology has been disappointingly slow in maturing to the point at which it can provide real relief in coping with increased levels of network complexity. The promise of network management systems and equipment is the hope that many of the processes now being conducted either manually or else somewhat inelegantly via the use of multiple incompatible network management systems can eventually be automated. This would allow IS/telecom managers the luxury of shifting the focus of their attention over to strategic applications and away from hardware.

The term network management is sufficiently ambiguous and broad in scope that it still manages to convey a wide range of meanings to different players in the communications industry. (It may even rival groupware for the dubious honor of being the semantic black hole from which explanations never emerge unscathed.) For present purposes, network management will be discussed in terms of stand-alone systems and equipment that track, monitor, and in some cases control the performance of both local- and wide-area networks. Such systems have become essential linchpins in increasingly complicated corporate networks. Their importance in the current environment cannot be overstated, because communications links are truly the lifeline of today's businesses. However, much has been expected of these systems and far less delivered with respect to the basic needs of the network manager. Until techniques such as artificial intelligence (AI) allow network management systems to become more sophisticated so that they can, in effect, allow networks to run themselves, the human process of network management will continue to focus on the necessity of performing these functions manually.

The current market for network management systems has been projected by one market research firm to reach almost $2 billion by 1995 [9]. Such healthy growth underscores the continuing need for these systems and the fact that they have already become commonplace fixtures in most corporate networks of any size. One of the precipitating factors for the widespread deployment of network management systems throughout private networks was divestiture; another was the rapid rise in the deployment of data communications networks. Before divestiture, AT&T managed circuit quality via their own internal facilities. AT&T's corporate customers often never knew when an actual network problem arose or, in some cases, that there even was a problem. If customers experienced circuit outages, they contacted their AT&T account representative, and the problem was addressed.

After divestiture, the picture changed dramatically. For one thing, on-site equipment that was increasingly necessary for data transmission (for example, smart hubs, modems, multiplexers) proliferated and became the responsibility of the company, not the local exchange carrier. As a result, the requirement for network management systems was created, and modem manufacturers were among the first to see both a need and an opportunity. The first major data communications network management systems, therefore, were designed to control, configure, and monitor modems. Eventually, it was *de rigueur* for all major modem manufacturers to come out with systems that monitored modem performance, often accomplished via a secondary channel apart from that over which the primary flow of data was carried. Such systems generally monitored the basic EIA signal activity in a network of modems and reported performance anomalies back to a group of network technicians stationed at a centrally based control center facility. These systems could also typically monitor line transmission quality to detect any degradation of signal or changes in such parameters as signal-to-noise ratio, impulse hits, line hits, and so on.

The purpose of such monitoring was originally quite simple and straightforward: to minimize the impact of outages occurring in the network. When a line or modem was not functioning properly, the device would so alert the operator. Later, as these modem-based systems became more sophisticated, they also took on additional functionality, such as the ability to adjust modem settings remotely or to switch or reconfigure circuits. Other important functions were eventually integrated into the systems, such as the ability to track historical performance characteristics or store information concerning equipment inventory and other physical network parameters.

From this simple and skeletal definition of network management arose an eventual proliferation of related systems and equipment, much of it geared toward a wide array of other devices including LANs, routers, intelligent hubs, PBXs, T1 multiplexers, and so on. As each of these systems evolved (as developed by individual vendors marketing the products associated with them), so did the capabilities and functionality of each. For their own purposes, many of these so-called element manager systems performed fairly sophisticated functions. In many cases, the basic hardware platforms used for these systems included stand-alone PCs, minicomputers, or workstations; or, in some cases, built-in functionality, as in the case of LAN management systems.

As network management systems for individualized equipment evolved, a particular irony began to surface. These systems solved the specific problems they were intended to address, but they did so in a proprietary fashion, so that one vendor's system was unable to communicate with another's (in some cases, this applied even to separate systems from a single vendor). Furthermore, they often used overlapping sets of hardware, which, from a user standpoint, was hardly cost-effective. Finally, in the long run, they served only to aggravate the larger and more comprehensive problem of network management by creating a "parallel universe" of competing and incompatible network management systems and products.

These developments set the stage for a new round of systems that were specifically designed to alleviate the multivendor network management problem. Early in this developmental phase, there were few available solutions. A handful of manufacturers developed multivendor systems that provided common operating interfaces between equipment from multiple vendors. Avant-Garde (now part of Boole and Babbage) was one of the first companies to offer this kind of capability.

In general, however, such approaches remained limited in their ability to accommodate the ever-growing variety of network management information that was being generated but not assimilated. And, ironically, every new "black box" that was introduced into the marketplace with the express intent of solving the incompatibility problem only added to that problem by creating yet another unintegrated system. The result was that the network control centers that monitored large corporate networks could often be found filled with row after row of consoles requiring varying degrees of attention from the network technicians and operators whose job it was to oversee them.

What was clearly needed were devices that could oversee all of the network management operations at a given facility, or at multiple facilities throughout an enterprise network—an approach that came to be known as *integrated network management*. However, the devices that perform these complex chores require significant amounts of processing power and must be compatible with most of the major corporate networks already deployed and operational. The task of developing these systems, therefore, naturally fell to the major computer makers and telecommunications providers: IBM, DEC, AT&T, British Telecom, Hewlett-Packard, and several others.

In 1986, IBM announced NetView, an SNA-driven host-based set of software tools that offered the capability to integrate and manage a number of other existing network management products, including IBM and non-IBM devices. The real importance of the announcement, however, was that the new system would become the focal point for an array of other network management systems through a single interface, thereby providing the so-called "single view of the network." NetView supported a number of other specific IBM products, including 3725 network control program, 3720 communications controllers, token ring LANs, and diagnostic modems. Later, IBM added additional multivendor capability to the product via a strategic alliance with Bytex, a maker of matrix switching equipment, and NET, a T1 multiplexer vendor. In addition, IBM had planned to use NetView as a means of tying in voice-related network management via its Rolm product line. Although this never took place (due to IBM's divestiture of Rolm), it nevertheless had significant implications for yet a higher level of network management—one that could conceivably handle voice as well as data.

Hard on IBM's heels came a number of other vendors, all purporting to capture the holy grail of integrated network management. Although oriented more to voice than data, AT&T's Unified Network Management Architecture (UNMA) posed a significant challenge to the IBM approach, with its ability to manage other AT&T network management subsystems (element managers). Digital Equipment Corporation announced its Enterprise Management Architecture (EMA), representing a less hierarchical, distributed peer-to-peer approach. DEC's strategy for developing multivendor interfaces was to provide application programming interface (API) specifications to a select number of vendors who then developed domain or element level products in conformance with them. Some of the vendors that participated in this program included Codex, StrataCom, and other companies that had developed proprietary midtier network management systems. End users enjoyed maximum benefit by interfacing with a high-level system like DEC's. Another major computer maker, Hewlett-Packard, entered the fray with its OpenView product, targeting both wide- and local-area networking and sharing certain similarities with the DEC approach. The HP approach was later endorsed by the Open Software Foundation (OSF) as a method that could be used as a model for future network management systems being developed to accommodate the growing deployment of so-called enterprise networks, based on linking LAN internetworks. This model is known as the Distributed Management Environment (DME).

FUTURE DIRECTIONS FOR NETWORK MANAGEMENT

One clear pattern for the long-term direction of network management lies in the integration of voice and data capability. Another is a shorter-term trend toward the ability of discrete, individually configured multivendor network management systems to communicate freely among themselves. IBM and DEC responded to both of these goals in 1991 when they announced a consolidation of their respective network management approaches. IBM unveiled SystemView and DEC announced Polycenter, both of which approaches consolidated management of computer resources with that of the network itself. Another significant development was an agreement between AT&T and IBM to link their respective network management systems—UNMA and NetView—thus for the first time creating a first-level network management approach that could integrate both voice and data.

Although such approaches might help alleviate the problem of multivendor network management in the short term, both users and the communications industry in general must in the long term look to the development of standards as the ultimate solution. OSI network management is expected to be an important factor here, along with emerging TCP/IP-based approaches for LANs and LAN internetworks. OSI, the seven-layer protocol stack that is expected to become a major industry standard for future open systems, is being fostered and developed at the network management level by the Network Management Forum, a group of major manufacturers collaborating on the development of individualized network management protocols such as CMIS and CMIP. The forum's basic charter is to develop protocols and interfaces that can be used on an industrywide basis. In the short term, however, TCP/IP-based approaches are gaining strength and prominence, primarily as the result of perhaps one of the biggest challenges in the network management arena to emerge in the early 1990s: the management of intelligent hubs, routers and bridges, and other types of devices specifically designed for the

LAN internetworking environment. With respect to TCP/IP, the network management agent involved is the *de facto* simple network management protocol (SNMP), which performs analogous functions to OSI CMIP.

Another important development in LAN management is the development of the management information base (MIB), a common database format specifically geared for network management and an essential element in the structure of SNMP. In addition, six vendors have cooperatively developed a standard called the RMON MIB, which takes the MIB concept a step further by adding a remote monitoring dimension, allowing agents distributed across an enterprise network to report to a central monitoring facility. The Internet Engineering Task Force issued RFC 1271 for the RMON MIB standard in November 1991. SNMP-based MIBs have been developed for token ring, token bus, Ethernet, FDDI, and X.25 products and systems. There are two elements of the standard being developed—MIB I and MIB II—and, as described earlier, the Internet, via a repository feature, is functioning as the central clearinghouse for users and vendors that require information or have the need to resolve problems concerning the standard.

As a capability representing the next generation of network management in increasingly LAN-dominated environments, the RMON MIB will advance SNMP agent capability an important step further. However, there are some who argue that the RMON MIB approach is too closely aligned with a centralized approach to network management that does not align itself properly with the new distributed peer-to-peer architectures that enterprise networking is headed toward. Furthermore, the eventual role of OSI in network management is far from settled. In particular, CMIS and CMIP protocols may eventually yield a new generation of products that can interoperate and exchange vital network management data even more deftly. However, regardless of the approaches adopted in the near future, the promise of the so-called "single view of the network" will continue to remain an elusive goal, especially as networks continue to decentralize.

Another interesting facet of network management is its continuing use by vendors as a major marketing hook. Equipment vendors have recognized that controlling the network management function provides a marketing gateway for account management (that is, the flow of products and services to the customer), especially because users are beginning to factor network management heavily into their buying decisions. For vendors, network management capability has become an item that, profitable or not, they must have in their product portfolio in order to keep up with customer demand.

A 1991 study on network management by the International Data Corporation (IDC) cited the emergence of several trends toward integrated network management. One major pattern was the evolution of a three-tiered paradigm for network management among large corporate users, as shown in Figure 6.1. The three tiers are as follows:

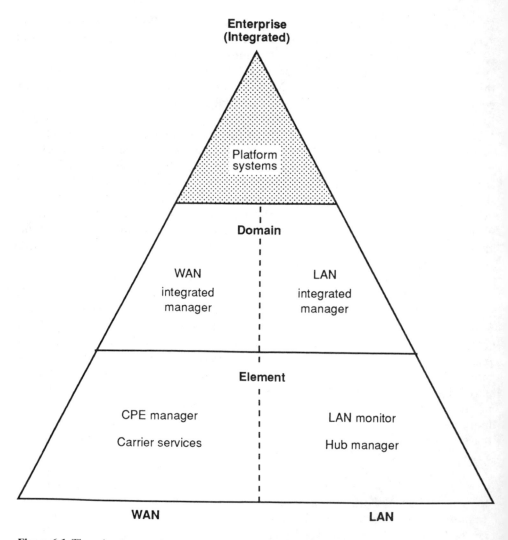

Figure 6.1 Three-level network management hierarchy. (Source: IDC 1992.)

- *Element managers*: the basic systems that monitor devices, including intelligent hubs, T1/T3 multiplexers, and PBXs;
- *Domain managers*: the next level in the hierarchy, which can be described as multiequipment systems that integrate information from various element managers;

- *Integrated network management systems*: "umbrella" systems that collect and aggregate traffic from all of these sources, representing the epitome of the centralized approach to network management (for instance, NetView and UNMA).

This three-tiered approach was largely the result of the evolutionary growth of various types of systems and equipment over the years, rather than a calculated plan on the part of either users or vendors. The first tier, element managers, have been in place for a number of years, growing in scope and number as more and more devices were added to the network. As these systems proliferated, so did the number of terminals that had to be monitored. Although it's not necessary to assume a one terminal–one person rule with respect to monitoring these devices, it's apparent that each system that was deployed created the need for at least the partial involvement of a staff person to periodically check for alarms and so on. The second level of network management, domain managers, then became a means of integrating information flow about network performance being culled from the various element managers in the network. Thus, somewhat in the same way that electrical power must be "stepped down" through a series of transformers, the information flow from a large number of different devices needs to be consolidated before it proceeds up the hierarchy, lest network operators become overwhelmed with data on network status.

Developing a hierarchical flow among tiered network management systems and the devices they monitor poses distinct challenges for the integrated systems at the top of the hierarchy, suggesting a chink in the armor of the centralization concept itself. For example, a major problem frequently encountered with the tiered approach is the phenomenon of cascading alarms. Imagine a network with three levels of multiplexers. If one of the lines at the lowest-level multiplexer were to fail, an alarm condition would register at that third-level multiplexer. This alarm condition might then be passed on to the second-level multiplexer, and then finally up to the first (master) device in the chain. Yet, to the operator looking at the first-level network management console, it might be impossible to pinpoint exactly where in the network the problem was occurring. As can easily be imagined, in actual practice, such conditions can become even more complicated, creating more problems than they solve. In order to address this problem, some vendors have developed products that can provide more sophisticated control and pinpointing as to where exactly in the network problems might be occurring and how severe they are in terms of their effect on the network considered as a whole. In general, however, the challenges posed for network management systems to ferret out this kind of information with real precision are daunting and clearly point to the value of AI and expert systems in eventually being able to intelligently sift through an overwhelming amount of irrelevant and possibly conflicting data.

All of the foregoing seems to suggest that there are theoretical and practical limits to how much information can reasonably be handled by a first-level network management system such as NetView or UNMA without the use of promising but still distant technologies such as AI. It also calls into question the efficacy of centralization as a model for the management of increasingly distributed systems. How much information can effectively be concentrated and delivered to a single operator about a huge and complex enterprise network with countless nodes deployed across any number of countries? A suitable analogy here might be attempting to drink from a fire hose—hence the need for intermediate-level systems to process the enormous amount of information about network activities being developed by lower-level systems.

Although, from a product development standpoint, AI capability has been extremely slow in coming, there are a number of positive signs. The first use of AI in test and measurement equipment occurred in 1991. Hewlett-Packard, long considered the "IBM of the test and measurement industry" introduced a protocol analyzer, a sophisticated device used primarily for troubleshooting data communications problems, which used AI heuristics for fault isolation purposes. In essence, the device electronically simulated the logic tree of evaluative judgments that a test and measurement technician might go through to identify a problem. However, testing a single data communications line using AI is one thing; managing a complex 10,000-node geographically dispersed network is another. Accordingly, most analysts don't expect AI capabilities to significantly come into play in the integrated network management arena until at least the next decade.

NETWORK MANAGEMENT OUTSOURCING

Up to this point, the primary focus has been on network management systems used by companies on their own premises to monitor their own networks. The emphasis has also been primarily on communications, rather than computing. In this scenario, the network management equipment must be purchased, deployed, and staffed with capable operators. There is considerable expense attached to this approach and, as discussed earlier, finding the proper staff to "baby-sit" these systems has become increasingly difficult. For example, in line with the outsourcing trend discussed earlier, there has been a move on the part of some companies to back away from the attempt to handle all these tasks internally. There are varying reasons for this trend, ranging from the cited inability to develop fully qualified in-house staff, to liability issues, to the basic economic advantages of one approach over another. Large corporate users generally have more of a tendency to move in the direction of maintaining their own facilities, whereas small- to medium-sized companies may not have the staff, the resources, or even the desire to take on this level of managerial responsibility. However, even among larger users, there are often incentives to outsource, depending on the specifics of any given situation.

For those companies that have chosen to outsource the network management aspects of their operations, there are several alternatives. A number of service providers now offer a range of network management services for the purpose of monitoring a customer's network on a remote basis. Generally, these services are available directly from carriers such as MCI and the RBOCs or from specialized third-party providers, such as AT&T's Paradyne subsidiary with their NetCare service, or General DataComm.

The specific outsourcing of such functions will here be referred to as network management outsourcing, so as to distinguish it from the outsourcing of IS or other telecommunications-related activities, such as systems integration or long-term planning [10]. Network management outsourcing rapidly gained favor among many companies in the early 1990s as they scrambled to reassess their networking needs and requirements. In 1990, for example, the Chevron Corporation outsourced the physical portion of their telecom network to AT&T, including all the attendant network management functions [11]. Other companies known to have taken similar approaches include Unisys and the Hyatt Corporation. Traditionally, one of the major drivers for this approach are the cost savings that that be achieved. In addition, there appears to be an increased tendency for companies to outsource their communications network management if they have already done so with their data center activities. All in all, the general demand for such approaches appears to be on the rise, although the full extent to which outsourcing will be relied upon is a subject for healthy debate. According to figures from the Yankee Group, the market for outsourced services will grow at a steady rate, becoming a more than $5-billion business by 1995. However, other market-watchers such as IDC estimate their eventual adoption by large corporate users to be considerably less.

A step beyond network management outsourcing, which, as we have defined it, refers to the performance monitoring of the physical network, is outsourcing the *entire* management of the network. This is an important but often confusing distinction, because monitoring circuits does not represent the same level of effort as taking over the full spectrum of management activities associated with planning, operating, and maintaining a network. For the purposes of this discussion, this will be referred to as network outsourcing. Examples of this approach have multiplied in the early 1990s, as companies such as J. P. Morgan and Unilever have turned over the running of their networks to major carriers. In the case of J. P. Morgan, British Telecom (BT) literally took over two out of three of their worldwide networks, a contract worth $20 million to BT but expected to save its client $12 million in network operating costs over a five-year period [12]. Carriers such as BT and AT&T are well positioned to provide this kind of capability to multinational corporations because they have very large global presence and the clout to develop locally based carrier facilities when needed. They also have developed major hubs, strategically positioned in key geographic locations, which act as concentration points for a company's international traffic. Much of the advantage provided by the carriers in these arrangements has to do with managing sheer network complexity. For example, Unilever, which handed over its network responsibilities to Sprint, has over 500 companies operating in 75 countries.

As outsourcing is considered as a management option, users need to evaluate to what extent the involvement of third parties in network operations is supportive of strategic communications. This is the single most important aspect of the outsourcing decision. Certainly once they're in place, turning over existing physical facilities (circuits and modems in the case of Chevron, for example) is not likely to impair the ability of a telecom department to achieve this objective. In fact, in some cases companies that chose to outsource some of the more routine functions within their organizations were motivated by the need to devote more time and attention to planning and other strategic considerations in network development. But moving beyond the issue of network management outsourcing and into the larger one of complete takeover of a network, the fundamental question remains: Can companies that have recognized the need to develop strategic uses for their telecommunications resources rely on outsiders who don't know their business as well as they do to come up with the right applications, or even creatively implement existing applications?

This networking conundrum, as framed by the public versus private network dynamic discussed in Chapter 5, will continue to be a key question for corporate IS/telecom management in the years ahead. The answer to the question, moreover, will likely depend on a variety of individual variables that each company must evaluate on a case-by-case basis. A company with major network growth and expansion ahead of it must evaluate the capability of its existing staff to take on the challenge. If existing staff is not equal to the task, it will have to weigh the costs of adding to staff against the costs of getting outside help. However, a lot more than just short-term costs are at stake. For example, an important consideration is the issue of the fostering and maintaining a sufficient knowledge base among one's own staff. Will those companies that do take the harder road and continue to utilize dedicated staff that can grow into a deeper understanding—not only of the basic telecommunications capabilities available via public and private networking, but also of *how those capabilities can specifically enhance and augment the basic corporate charter of their company*—develop a sustainable competitive advantage over others that decide to sign away this responsibility to outside developers? Alternatively, would the use of management consultants and outside experts who have developed their own experience base, "cross-pollinated" by a wider exposure to state-of-the-art practices being implemented across many different vertical markets be able to, with their larger industry view, bring more innovative approaches to the table?

The question of how increasing departmental control of networking resources in the context of decentralized and downsized systems applications will affect IS/telecom corporate management is far from settled. IS/telecom managers should keep in mind that two factors may converge to keep this trend alive and well throughout the 1990s. First, the commoditization of networking products (as discussed in Chapter 2) will create more "plug-and-play" options for departments to build "instant networks" with immediate global scope and reach. This will be reinforced by the development of a new generation of downsized switches, based on the convergence of T1 multiplexer, intelligent hub, and router capability in a single device. This technological trend, in conjunction with the long-awaited incorporation of data networking technologies into the public network (frame relay, B-ISDN, SMDS, and IP routing) will eventually simplify the process of allowing individual departments to "plug into" the corporate network utilities discussed earlier.

Finally, a dominant theme of this book has been the recognition that strategic applications will be developed by and for users, at the experimental laboratory called the desktop. For IS/telecom managers, this has several implications. First, it suggests an organizational reorientation that will take this fact into account. Second, it implies the need for a new definition of the kind of expertise required to provide the necessary support for such applications development. IS/telecom managers will have to become more attuned to the needs of end users and learn to act as network resource facilitators, in the attempt to meet each department's individualized sets of requirements for enterprise networking. They must, in any event, avoid the mistakes of the past in terms of serving the needs of their "internal customers," the end user. Ideally, such professionals will have developed a keen sense of how basic business tasks are approached and accomplished, and how the flow of information can expedite, inform, impede, and otherwise affect these processes.

NOTES AND REFERENCES

[1] "It's Reality Time," *Computerworld*, 29 April 1991, 81.
[2] W. Eckerson, "Firms Empower Business Managers to Fill CIO Role," *Network World*, 25 March 1991.
[3] This breakdown is based on information obtained from the Communications Managers Association (CMA).
[4] The survey cited was performed by *Network World*, a communications trade publication. The results were based on a sample of 622 randomly selected subscribers to the publication, out of a total mailing of 2000. See "Network World Salary Survey," *Network World*, 28 January 1991.
[5] "*Telecommunications: The Decision Makers*," Wall Street Journal survey of International Communications Association (ICA) members, May 1990.
[6] *Ibid.*, 3

[7] J. Chiaramonte and J. Budwey, "Finding and Keeping Good People," *Telecommunications*, February 1991, 55.

[8] See "Outsourcing and Centralization: Another IS 'Tug of War'?" *Telecommunications*, November, 1990.

[9] Estimates are from the Market Intelligence Research Corporation (MIRC).

[10] There is a fair amount of confusion in the industry over the definition of the term outsourcing. In some contexts, the term network outsourcing has been used to refer to outsourcing the entire operation of the network, rather than just the performance monitoring aspects related to the physical network. Accordingly, I've tried to draw some distinctions here that I hope will alleviate some of the confusion.

[11] *Information Week*, 25 February 1991.

[12] See John Keller, "More Firms Outsource Data Networks," *The Wall Street Journal*, 11 March 1992.

Chapter 7
Case Studies in Strategic Networking

Companies attempt to impose order on information by designing computerized management information systems. Some of these, it turns out, are intended to buttress the old system by employing computer and communications links merely to expand the cubbyhole and the capacity of the communications channels. Others are truly revolutionary in intent. They seek to crush the cubbyholes-and-channels system and replace it with free flow information.

—Alvin Toffler

According to some industry observers, writing about the strategic use of communications is an unusually difficult task. I'm inclined to agree. The reasoning behind this observation is fairly simple: companies that have managed to implement such applications successfully are often reluctant to publicize them. However, although there is a great deal of truth to this generalization, there is another side of the coin to be considered. Generally, the confidentiality caveat is especially relevant when it comes to big-systems implementations of the sort ardently discussed in the trade and business press. Classical examples include American Hospital Supply's use of computers for on-site ordering and inventory and American Airlines' SABRE system. Such large-scale projects are indeed best developed stealthily, as is the case with any other corporate undertaking that requires an element of market surprise. Then, at such time when the full impact of the project has managed to deliver a preemptive first strike and has become a major marketplace success, can the stories behind their development be circulated among a wider audience—one that, of course, includes a company's competitors.

For every one of these big-systems stories, however, there are thousands of others that deal with the ways in which telecommunications has had perhaps less dramatic but equally effective impacts on a company's marketing, sales, and support activities. Although the big-systems stories are especially useful as attention-getting, leading-edge indicators of major trends, they by no means tell the whole story. Ultimately, the complete picture of the successful application of computer and communications technology is best presented as a mosaic of many individual transformations taking place throughout a variety of commercial, corporate, and government environments.

The intent of this chapter is to provide a snapshot in time of this mosaic and a useful sampling of applications, which will, via the process of extrapolation, offer the reader an intuitive sense of the larger picture. I have not attempted to provide comprehensive documentation of strategic implementations nor to dwell on any one application with the kind of depth that might be the province of original research. Rather, the intent is to provide a wide-angle view of how telecommunications is being applied successfully across a broad spectrum of vertical market industries and how those applications are changing the way that companies conduct their business in very fundamental ways.

There are certain basic themes that cut across such strategic applications and how they contribute to business profitability, some of which were discussed in Chapter 3. These include (but are not limited to) the capability to:

- Increase the scope and scale of business markets via the advantages afforded by time and distance insensitivity;
- Improve direct channels of communication between companies and their customers, subsidiaries, suppliers, and strategic partners;
- Bring new products and services to market more quickly and efficiently and to custom-tailor them to specific customer requirements;
- Improve internal communications among in-house departments, resulting in the ability of the organization to move more quickly and adapt to changing conditions in a highly competitive, real-time-oriented global marketplace;
- Provide better responsiveness to existing accounts and customers in terms of after-market sales and support, paving the way for repeat sales and better customer relations;
- Respond, in service-oriented organizations, to customer needs more quickly and effectively by externalizing the knowledge base available among company representatives;
- Reorient the organization toward providing solutions to customers and accounts with specific needs to fill rather than simply offering cookie-cutter products that must be "matched" with the appropriate consumer;
- Identify markets and targets of opportunity for sales, effectively allowing the organization to find the customer rather than leaving the process to happenstance or random logistics.

Clearly, there are critical dependencies and interrelationships between these items.

The remainder of this chapter will provide a series of descriptions of strategic applications throughout a variety of vertical industries and technologies. The attempt has been made to group these applications under many of the above categories, but areas discussed in Chapter 3 are also included where appropriate.

STRATEGIC APPLICATIONS IN MANUFACTURING

Computer integrated manufacturing (CIM) involves the use of computer and communications technology to streamline and automate processes that may or may not have been previously under some other form of machine-based control. Many manufacturing operations are centered on the use of Ethernet LANs, which in some ways are more suitable for manufacturing applications than token ring. In addition, special token-ring-oriented protocols, such as the manufacturing automation protocol (MAP), have been developed over the years to foster OSI-based open connectivity between multivendor systems, although progress in this arena has been disappointing when matched against user expectations. CIM has become commonplace in many corporations and is being used to great advantage to create high-quality, custom-made, customer-responsive products. Such products give U.S. manufacturers the ability to compete better—especially in an area in which considerable ground has been lost to offshore enterprises that have the advantage of cheaper available labor forces.

New and interesting approaches to computer-based manufacturing, however, are now being widely discussed and implemented by U.S. companies anxious to regain this lost ground. One initiative involved a group of the nation's top business leaders who convened in the summer of 1990 to discuss and brainstorm more forward-thinking approaches—steps that they felt American manufacturers needed to take to stay competitive in a rapidly changing global marketplace. Lehigh University was the sponsor and prime mover for the study. The executives were drawn from major U.S. multinational corporations, including AT&T, Motorola, IBM, General Electric, and Boeing. The results of this series of meetings were documented in a two-volume report, *21st Century Manufacturing Enterprise Strategy*, described by *Business Week* as "a daring blueprint that . . . could enable U.S. industry to match—and surpass—the ambitious programs under way in Japan and Europe to develop tomorrow's factories" [1].

Perhaps the most interesting—and controversial—concept to come out of the study was the idea that, with an extensive company-to-company, nationwide computer network in place, new task-oriented "virtual ventures" could be developed to go after fast-moving markets. CIM itself is, of course, a major precondition to this kind of approach, but another key is easily reconfigurable and network-based production capabilities. According to the article, ". . . the network would facilitate concurrent engineering—developing a product as a collaboration of the design, production, marketing, purchasing, and service departments—either within the same company or among corporate partners. The network would also foster collaboration on 'virtual ventures' created for one project and then disbanded."

It remains to be seen whether U.S. industry will respond to this challenge. There is a fair amount of skepticism on this subject, especially regarding the likelihood of garnering the appropriate multiparty cooperation required for these virtual ventures. Nevertheless, it's likely that such concepts of computer network–based "agile manufacturing" are indeed the direction in which many multinationals are headed. Furthermore, it will be interesting to see what role development of the U.S.-based Internet/NREN network might eventually play in making the kind of recommendations put forth in the report a reality.

A Unique "Virtual Venture" Network in the Southeastern United States

Although the idea of "virtual ventures" might seem farfetched, there are already examples of projects underway that illustrate the enormous potential inherent in the concept. In some cases, these examples include the participation and involvement of federal, state, and local government, as well as the university community. At the heart of these new business-education-government partnerships is the technology of computer networking.

In South Carolina, a unique network has been established that links manufacturers with technical experts in the state's universities and colleges [2]. The Southeast Manufacturing Network, or SemNet, is a virtual network running on the Internet regional SURAnet, with connections in at least a dozen southeastern states in addition to South Carolina. The network currently links the following organizations:

- General Motors Corporation
- Digital Equipment Corporation
- IBM
- United Technologies Corporation
- Kendall Square Research Corporation
- The High Performance Manufacturing Consortium
- The Southeast Manufacturing Technology Center (SMTC)
- The University of South Carolina
- The South Carolina Research Authority
- 370 South Carolina manufacturers
- 16 technical colleges in South Carolina

The current backbone capacity of the network is 1.5 Mbps, with a future 45 Mbps upgrade planned. The additional network capacity will be used to support applications such as interactive videoconferencing and the capability to exchange three-dimensional images. Network subscribers can also offload CAD/CAM applications to a remote DEC VAX vector processor via the network.

The benefits to members of the network consortium are significant. First of all, it allows small- and medium-sized corporations to become part of a larger nexus of suppliers and fabricators, all operating on a cooperative basis. Second, it allows economies of scope and scale to be introduced via linking the collective abilities of separate entities. Third, for suppliers working collaboratively on the network, it provides the electronic equivalent of just-in-time manufacturing.

Perhaps the most interesting aspect of the new government-supported network is the development of what the consortium calls an "electronic bidding board" under the auspices of the Department of Defense (DOD). Under this arrangement, if DOD requires a quote for parts shipments, it posts the information electronically. In turn, potential suppliers then download specifications and drawings for evaluation. The objectives are to create a level playing field for possible providers, shorten provisioning cycles, and reduce overall costs of procurement for all parties. More importantly, eventually the bidding board will be used for non-DOD-related efforts.

Using CIM to Streamline Manufacturing Processes

Corning Asahi is a Pennsylvania-based company, a joint venture between Corning, Inc, and Japan's Asahi Glass Co. The company, which manufactures video products, ran into serious financial problems during the 1980s. In 1985, it began implementing CIM network systems and architecture to enhance its capabilities in the production of television picture tubes. The changes were the result of a five-year plan that had the objective of cutting manufacturing costs by $50 million and significantly increasing the basic quality of finished products. At that time, the company's computer resources consisted mainly of a single DEC VAX used to control operation of the company's melting furnaces. As a result of the new approach, much of the company's manufacturing operations are now computer-controlled. The results of this effort to date have reportedly made the company far more competitive and returned it to profitability. For example, CIM has reduced the number of specific manufacturing steps taken in the production process from 265 to 215. In addition, since 1985, the operation has tripled manufacturing output and reduced manufacturing costs by as much as 22%. According to the company, of that 22%, 65% of the cost savings resulted directly from improvements in the manufacturing process, 15% from automation of procedures, and 11% as the result of producing fewer defective parts [3].

OPTIMIZING INTERNAL COMMUNICATIONS: GROUPWARE IN ACTION

As discussed in Chapter 4, the definition of groupware is somewhat amorphous. As embodied in leading-edge products such as Lotus Notes, it points the way toward a time when workgroup computing software may represent the very essence of corporate departmental activity. In the not-too-distant future, professionals may never enter or leave the office without first "checking their groupware."

In theory, groupware can function as a kind of "living document," providing an ongoing record of a business's major internal and external transactions, which can enhance accountability and foster task-based group cooperation both directly and indirectly. In the example that follows, however, groupware has been used in an unquestionably unique fashion, pointing to a future direction that some might view as intriguing, and others disturbing. In the application to be described, groupware has been used to place the activities of all corporate employees and operations into a kind of "virtual fishbowl." Every action is computer-recorded and hence known to a wide group of individuals throughout the corporation. In this system, those individuals and groups that do not hold up their end of the process become immediately singled out and identified by senior management. Whether this kind of approach represents the wave of the future remains to be seen. However, it vividly illustrates the potential for computers to function as management watchdogs and to increase levels of accountability, regardless of whether such approaches are ultimately judged to represent the basis of sound management practice.

Cypress Semiconductor: Using Groupware to Track Performance

The following application of groupware might seem unbelievable if it weren't true. California-based Cypress Semiconductor has created a novel use of groupware that constitutes the very essence of its competitive strategy, with respect to both its own employees and external suppliers. The company mandates companywide use of open groupware systems, widely accessible among different departments in the company. Company sources attribute Cypress's counterrecessionary levels of success during the early 1990s to the use of the system.

Here's how it works. The basic idea involves a computer-based application of the concept of management by objectives. The objectives in this case are those set by employees themselves in the course of their normal responsibilities and annotated via computer with the expectation that the milestones set as goals will be fulfilled. Once entered, the information is sent to a central database where senior-level managers review it with an eye to overall corporate objectives. This in and of itself is considerably more innovative than the way most companies operate in terms of their use of computer resources. But Cypress has taken the entire operation described thus far a bold step further. (Readers can judge for themselves whether such an approach is really the wave of the future or heralds the potential for a new kind of "totalitarian" computer-based control of the workplace.)

The key to Cypress's approach is what happens when deadlines and performance objectives are not met. When this happens, a custom-designed program that has been described as "killer software" is triggered:

> The basic idea behind killer software is to guarantee that departments operate at peak efficiency by forcing line managers to make good decisions—or else it shuts them down. Like a hydra-headed gatekeeper, killer software prowls through Cypress's computers taking notes on moment to moment performance in each department. When the software finds a unit that's slipping behind schedule, it simply turns all computers off there. [4]

Draconian? Dickensian? Or smart business? Whatever the judgment, the approach appears to have worked for Cypress, which was reporting increases in net revenue of as high as 32% at a time when the rest of the semiconductor industry—including industry giant Motorola—had been languishing in the shadows of recession. The program performs other functions as well, such as scanning inventory to make sure that parts don't sit unused beyond a predetermined timeframe. This particular feature has reportedly enabled the company to cut inventory holdover to ten days from cycles that previously ran over a year in length (in some cases approaching almost two years). The program also applies to the company's outside component suppliers. If shipments of a particular stock item are late, and the company hasn't been notified about the late shipment, the transport company will be turned away at the door when it finally shows up. The company must then work out a detailed plan to ensure prompt delivery in the future, at which time the embargo is removed.

IMPROVING COMMUNICATIONS WITH CUSTOMERS AND SUPPLIERS

Electronic data interchange (EDI) is an important standard for the exchange of transaction-oriented information between companies. Typically, this information relates to the purchase or delivery of products and services. Most importantly, EDI allows billing and ordering data between companies and their suppliers and customers to be exchanged electronically rather than in paper form, often with a resultant annual savings of millions of dollars. Thus, invoices, specifications, material requisitions, shipping schedules, receipts, and inventory data are typically exchanged via EDI links between companies. The standard itself is being developed under the auspices of ANSI's X.12 committee but is firm enough to be found in use among a substantial number of U.S. companies.

EDI allows the new Reichian "enterprise webs"—close alliances and partnerships between nationally and internationally deployed companies—to communicate quickly, effectively, and in a common format easily understood by all parties. The technology is now well established within the Fortune 1000; companies such as Sears and Ford Motor Company have mandated its use. Ford, for example, requires close to sixty of its suppliers to use EDI as a precondition of doing business with them [5]. Other large corporations in the process of implementing EDI conversion programs include JC Penney, Wal-Mart, K-Mart, Burlington Industries, and Levi Strauss.

How Sears Moved Its Supplier Base Toward EDI

In mid-1989, faced with the increased competitive pressures characteristic of the retail industry, Sears decided to abandon its proprietary approach to EDI and adopt the industry standard [6]. As a part of this reorientation, the company delivered the message to their more than 6000 suppliers: EDI would represent their preferred way of doing business. This aggressive EDI plan—reportedly one of the largest ever implemented—was to take three years, with completion targeted for the end of 1992. At a total cost of $5 million, the project was designed to occur in three phases. The first involved swinging some 350 catalog suppliers over to the standard. In 1991, Sears launched the second phase, which brought the remainder of its suppliers into the EDI mode, thus allowing electronic ordering and invoicing to take place. All in all, the effort represented an expensive and difficult proposition and many suppliers were hesitant to make the transition according to Sear's strict marching orders. However, the implementation plan appears to have been largely successful in establishing the levels of EDI-compliance Sears was looking for. In order to expedite the transition, the company offered incentives, such as free software and training.

Use of EDI in the Transportation Industry

The need for increased technological sophistication in communications has become evident even in such traditional smokestack industries as trucking and transportation. Just as major retailers have been telling suppliers to become EDI-compliant or risk losing competitive ground, they have also been delivering a similar message to their transportation partners. The benefits for retailers involve not only faster delivery and turnaround, but also the capability to track the progress of shipments far more accurately. Both the retail and trucking industries have been using EDI routinely for many years. In fact, at one point during the early 1970s, the trucking industry was considered a leading-edge developer of EDI applications. In terms of both these and other industries involved in such implementation, a major thrust in recent years has been an emphasis on the use of more standardized EDI codes.

With respect to the use of EDI in the transportation industry, a major buzz-word that makes the rounds is "quick response." In fact, a number of major retailers such as Levi Strauss and K-mart have quick response programs in place within the larger sphere of their existing EDI programs. The quick response approach is to retail what just-in-time techniques are to manufacturing—namely, an attempt to operate in a more efficient, market-responsive manner, eliminating the need for unusable levels of stored inventory. The basic objective is to reduce the amount of cycle time involved in the process of stocking orders for distribution. In the case of K-mart, for example, a target goal has been to reduce the time from the cutting of a purchase order to delivery of merchandise from two to three weeks to five to nine days. The challenge for the truckers, based on this requirement, has been to cut actual road time involving a delivery from three days to one day. This entails changing a fair number of traditional practices and adopting state-of-the-art bar-coding techniques used in conjunction with ANSI X.12-based EDI technology. Other changes required are less technology-related but pertain to developing increased inter- and intracompany cooperation between trucking firms. However, despite early protests by some segments of the trucking industry, many of the changes have allowed as much as an estimated 40% of unused capacity to be more fully engaged [7].

Using Electronic Information Services to Increase Transportation Efficiency

One of the most interesting applications for matching customer requirements and available transportation capacity involves the use of electronic information services. Through a publicly available videotex system or BBS, for example, individuals or companies with goods requiring shipment can post their requirements on a public videotex system that any transport firm in that serving area could access. The company that has transport resources most easily and quickly deployable to the customer site (within the cost parameters specified) could then respond to the request. Similar computerized systems have already been in use in the taxi industry for several years, allowing more efficent customer response. The same principle is easily generalized to the trucking industry. In fact, such a system has been operating for several years in France via their widely used Minitel system; any one of its 6 million users can post a transport request on the BBS and get a response from subscribing transport companies.

The potential for such individualized, customized types of transport services is significant. For example, if, via the RBOCs or some other information services provider, a similar type of public BBS system were to be adopted in regions of the United States, then transport and the delivery of retail services to the home might eventually become a reality. The increasing use of residential delivery services observed during the 1980s might be seen as the first steps toward this scenario. The most obvious example was Domino's Pizza, which grew phenomenally in a relatively short period of time on the strength of its policy of 30-minute home delivery. As electronic BBSs are deployed in the home, it's not too difficult to imagine locally based transport services that might be engaged to deliver a variety of retail items ranging from supermarket goods to stationery to home videos (assuming they are not delivered electronically by that time). In fact, in one interesting scenario, local merchants might plausibly underwrite the cost of providing the BBS terminals in exchange for having their goods and services listed in a first-tier data base available to local subscribers. It will be interesting to see how, in the years ahead, this trend takes shape and what new services might be developed.

Halliburton: Using Videoconferencing for Competitive Advantage

In the early 1990s, videoconferencing came of age and became widely accepted in many corporations as an alternative to corporate travel and a means of collaborative teaming. System cost was a significant gating factor. In 1991, one of the major videoconferencing providers, PictureTel, began offering an entry-level system for under $20,000. Other developments that spurred its acceptance included the advent of new switched digital services available from AT&T, MCI, Sprint, and other long-distance carriers. Offerings such as AT&T's Switched 384 are ideally suited for videoconferencing applications and will continue to pave the way for increased deployment, as will the availability of new customer premise equipment such as inverse multiplexers, which allow users to optimize bandwidth needed for such services. (An anecdote that is indicative of the change in corporate thinking concerning videoconferencing is that of a product manager with Northern Telecom who was required to fly from his corporate offices in Research Triangle Park, North Carolina, to another location nearby for the sole purpose of attending a major staff videoconference session!)

The Halliburton Company is an an oil and gas provider based in Arlington, Texas. In 1992, the company began a major upgrade of its existing videoconferencing systems, with the objective of improving customer relations and cutting costs. There was an interesting twist, however, to the company's decision to do so. Unlike other companies that delve into videoconferencing, Halliburton specifically intended to use it to gain competitive advantage.

The company installed the PictureTel systems in 30 of its worldwide locations, including an international facility based in London and domestic sites in Texas, Washington, California, and Alabama. The current network arrangement uses

AT&T's Accunet Switched 56 service for noncompany locations, but the company has a private T1 network in place for its own facilities. It will, of course, use the systems for internal communications. More significantly, however, the major impetus in deploying the system was to establish better communications with customers. Many reportedly are already equipped with compatible PictureTel systems; however, for those that aren't, the company will furnish them as loaners. In the oil and gas business, the systems are particularly useful in the supplier-customer relationship, in which the need often arises to communicate site drilling and geographical information on a near-real-time basis. With this information available, the customer's geological experts can expeditiously make a "drill or no drill" decision. A quick decision has a twofold benefit. First, drilling can begin immediately, minimizing the prospect of having equipment in the idle mode before the decision is made. Second, the customer can avoid the time and cost associated with sending its experts out to the actual well site [8].

IMPROVING SERVICE DELIVERY EFFICIENCIES

UPS Overhauls Its Core Business Using Advanced Information Technology

One theme seems clear throughout the varying threads of how networking has affected American business: companies that have learned to become successful in the information age—regardless of whether they were part of the so-called smokestack economy—were the companies that learned in one fashion or another to adapt to the new realities inherent in networking and communications. Longevity in American business is neither a *de facto* disqualifier nor guarantor. Western Union, for example, is a prime example of a company that has been in the electronic communications business for years but lost out on the promise of the information age because of its failure to adapt. On the other hand, information technology has transformed many older businesses whose management had the foresight to embrace new technologies as a means of either enhancing existing business or developing new markets.

An example of an older company that has learned new computer and communications tricks is United Parcel Service (UPS). According to company mythology, the UPS was founded in 1907 "in a closet sized basement office by a group of teenagers with two bicycles, one phone, and five month's rent." In 1991, UPS unveiled a new multimillion-dollar Information Services Headquarters and Data Center. This new facility symbolizes the importance of information technology to a company whose business consists of something as basic as delivering packages to a wide array of customers, but doing it as efficiently as possible.

A major objective in building the center was to move the company a step closer to paperless transactions. Thus, during a four-year period (1987–1991), the company dramatically increased its IS staff from 100 to 1700 employees nationwide. The capabilities afforded by this kind of commitment to new technology include such enhancements as Delivery Confirmation Service, for customers who require immediate delivery verification, including signatures, and the Package Tracking System. At the front lines, UPS drivers are equipped with hand-held computers that capture all pickup and delivery information, including an image of the customer's signature. At the end of the day, the system automatically transfers this information to a centrally based computer that keeps tabs on all major shipment activity throughout UPS customer serving areas.

Wireless Communications: Better Service for Avis Customers

In the package distribution business, the name of the game is to track the movement of shipments throughout the nexus of the transportation system. Other businesses have found that communications technology can be similarly used as "sensors" that allow a command center to pinpoint the status of each fleet resource at any given moment. For example, Avis, the rental car giant, used a radio-based network from ARDIS—a joint venture between IBM and Motorola—to develop a new check-out system available in fifteen cities.

The new approach involves the use of Motorola's wireless computer terminals made available to the drivers on Avis airport shuttle buses. The system allows prequalified renters to be taken directly from the airline terminal to their cars, skipping the check-out desk entirely. When the customer boards the shuttle, the driver matches the customer's names with an on-board manifest, then keys an identification number into the terminal. The data is passed through the ARDIS network to a mainframe based in Garden City, New Jersey. The mainframe, which stores a predetermined rental agreement, then instructs a computer terminal at the final rental location to print the appropriate agreement. The shuttle driver is then advised of the car's location, where the rental agreement will be waiting inside the car for the customer. Customers thus bypass the check-in desk entirely and need only show their driver's license to the attendant before driving off the lot.

Ethernet LAN-based Imaging at Kenwood USA

The subject of office automation and the paperless office was widely discussed in the late 1980s, but little in the way of applications surfaced. When a few major companies started making major moves toward this technology, however, many users began looking more seriously at the technology. American Express, for example, undertook a major conversion to the use of imaging for processing financial transactions, along with a number of other leading-edge adopters. In a sign of the potential importance of this new technology, Wang made a high-risk change in an attempt to jump into this market early, committing significant corporate resources to the project. There are now a wide variety of imaging companies in this promising new market segment, including such players as FileNet, Optika Imaging Systems, and all of the major computer makers.

Imaging and communications are heavily interdependent, because the support structure required for imaging is extremely bandwidth-dependent. Imaging systems and multimedia applications will also act as major drivers for LAN technologies such as FDDI over both fiber and unshielded twisted pair wiring, and ATM-based approaches that have rapidly been embraced by router, T1 multiplexer, and intelligent hub vendors working on the development of next generation products.

Kenwood USA is a California-based consumer electronics company that has implemented imaging throughout a number of its customer-service departments. The communications support structure for the system involves Ethernet over unshielded twisted pair (10BaseT) in conjunction with a Novell network operating system [9]. The changeover cost the company approximately $350,000 but is expected to save more than $100,000 per year in terms of significantly diminished requirements for warehouse space and the ability to far more efficiently respond to inquiries and communications from both dealers and customers. On this basis, the new system yielded an expected payback of 3.5 years. One internal user of the system was the warranty claims department, which on a typical day could receive more than 1200 pages of paperwork from its network of dealers. Under the old approach, information from incoming forms had to be manually keyed into a terminal connected to the company's mainframe. Conducted in this manner, the process was cumbersome and inefficient, because the original form could not be easily located by anyone wishing to check on status or expedite the process.

The imaging system that runs over the 10BaseT network infrastructure is FilePower, developed by Optika Imaging Systems. On the Ethernet LAN are the following nodes:

- A file server connected to an optical jukebox;
- Another server that supports a document scanner;
- PCs running the FilePower software;
- A server that negotiates mainframe traffic;
- Print servers.

Novell NetWare 3.11 was the LAN network operating system software employed at the time of the transition. A series of bridges are used to link the systems with other departments in the company that have requirements to share the imaged data base.

APPLICATIONS THAT ENHANCE CUSTOMER ACCESS

Voice Processing: Improved Customer Access at the U.S. Post Office

In conjunction with the use of computer data bases, telecommunications can be used to create a knowledge base that exceeds the capabilities of individual workers by orders of magnitude. Such an approach has been adopted by the U.S. Postal Service. Even the most dedicated civil servant is unlikely to possess knowledge of an operation that is so sweeping or broad that any type of customer inquiry could be accommodated. When customers call in to their local post office, there is unfortunately a random quality to the inquiry. They must hope that the individual fielding the question happens to be knowledgeable about their particular question. Furthermore, it's unreasonable to expect any one individual to be conversant in all facets of a given operation. But what if the range and diversity of questions that customers typically put forth via telephone could be anticipated, thoroughly researched, and then committed to a computerized knowledge base that could be accessed by each caller? This is essentially the approach taken by the U.S. Post Office for what they call their "Automated Telephone Information Service." The service is now operational in a number of major U.S. cities and uses audiotex to provide a menu of messages that callers can easily access.

What is perhaps most interesting about this approach is that it can reverse traditional expectations concerning the relative merits of the telephone versus an in-person visit to a facility. Thus, the quality of information (QOI) that the caller receives is probably equal to or even better than that available by a personal visit. In the case of the automated system, the human knowledge base has been codified and standardized and thus is not subject to the unpredictability of a random visit or phone call. The use of the system, therefore, offers a number of benefits to both postal workers and customers. For the postal service, it reduces the interruptions and necessity of dealing with callers on a case-by-case basis. For the customer, it enhances QOI, increases the likelihood of getting the information needed, and makes the service available on a 24-hour basis.

800 Number Service for Home Business Environments

Customer access to business has been changed dramatically by the impressive rise in the use of 800 numbers. The use of 800 numbers sends a simple and direct message to potential and existing customers: *we want to hear from you*. Furthermore, 800 numbers allow the organizational receiving point to be structured specifically for the nature of the call. Certain customers segments, for example, can be allocated special 800 numbers. The benefit is that call segmentation becomes an automatic process, reducing unnecessary hunt-and-search time on the part of both a company's employees and customers. 800 numbers, seen from one point of view as the cutting edge of "free communications," are now widely used in small- and medium-sized businesses to increase the scope and reach of their market presence. In one application, described below, they are also being adopted in record numbers by "homeworkers."

One of the first telephone companies to adopt an innovative new service—800 numbers for the use of professionals working at home—was Telecom*USA (now merged with MCI). Other carriers have followed suit. According to information from Pacific Bell, nearly 30% of all households in California include someone who works at home or runs a business at home. In early 1991, Pacific Bell began offering custom 800 numbers to homeworkers. Pac Bell offers the service to any customer with a residential phone line as long as the phone is used primarily for personal calls. The number costs approximately $5 per month and subscribers pay for usage ranging from $0.10 to $0.18 per minute. If the home phone is used primarily for business calls, then business rates go into effect. Pacific Bell will handle all custom 800 calls originating from any of its 10 regional service areas. However, in tandem with a long-distance carrier, arrangements can be made to extend the service to other locations outside of BOC's territory.

ACCESSING REMOTE MARKETS AND LABOR RESOURCES

How Fortune 1000 Companies Access Alternative Labor Markets

An increasingly important use of telecommunications by major multinational corporations is the ability it affords them to gain access to remote markets and business resources. Creating sustainable competitive advantage via access to offshore and remote U.S. labor markets is an approach that has been adopted by a number of companies. Citibank, for example, moved its data-processing facilities to Sioux Falls, South Dakota, in the 1980s. Data entry is a particularly attractive area for employing labor forces in various locations around the globe—groups of workers that can reliably and cost-effectively provide such services on an ongoing basis. In 1990, for example, keypunch operators in the United States were making approximately $6.50 per hour. However, there were areas outside of the United States where this work could be carried out for a considerably lower hourly rate. As a result, many U.S. corporations sought to take advantage of this disparity. In 1990, American Airlines began using labor forces in such locations as Barbados to process used airline tickets. The tickets are flown in daily to the off-site workers, who then key the information into terminals; from those access nodes, the data is then transferred to a mainframe at American Airlines headquarters in Dallas. Another example of this trend is an application developed by health insurance provider Blue Cross/Blue Shield, who contracts with various labor groups in the Farm Belt to perform similar routine data entry involving claims processing. And publishing giant R. R. Donnelly sends entire manuscripts to Barbados to be entered into computers prior to being printed and distributed [10].

NOTES AND REFERENCES

[1] "This is What the U.S. Must Do to Stay Competitive," *Business Week*, 16 December 1991, 92.
[2] Gary Anthes, "Small Firms Unite Through Net," *Computerworld*, 20 January 1992, 59
[3] This case study is a summary of a narrative that originally appeared in W. Eckerson, "Corning Asahi Turns the Corner Thanks to CIM Plan," *Network World*, 3 September 1990.
[4] Stephen Govoni, "License to Kill," *Information Week*, 6 January 1992, 22. I recommend this article as an interesting glimpse of the possible "dark side" of computers in the workplace.
[5] Alvin Toffler, *Powershift: Knowledge, Wealth, and Violence at the Edge of the 21st Century*, Bantam Books, New York, 1990, p. 121
[6] *Information Week*, 1 April 1991, 12–13.
[7] Douglas Bartholomew, "Grabbing the Wheel," *Information Week*, 28 October 1991, 26.
[8] J. Brown, "Firm Seeks Market Edge With Video Conferencing," *Network World*, 1 October 1990.
[9] K. Gilooly, "Firm Goes Paperless With Image Systems," *Network World*, 23 December 1991, p. 13.
[10] The source for American Airlines and R. R. Donnelly is Robert B. Reich, *The Work of Nations*, Knopf, New York, 1991, p. 211.

Chapter 8
Future Prospects and Imaginable Scenarios

It's not change, it's the rate of change.

—*Marshall McLuhan*

The future of computer communications and its strategic use throughout a wide variety of vertical markets is a wide open prospect at this juncture. As mature as specific technologies appear to be in many cases, the industry is still at the ramp-up phase of a steep logarithmic curve, with many areas—such as internetworking—very much in the early stages of development. The next important phase will center on a significant range of user-oriented applications, beckoning those companies and individuals with the requisite imagination and vision to take on the challenge.

Computer networking will be a critical facilitator of the increasingly agile corporate organizational structures needed to compete in a world market driven by instantaneous feedback and constant change. Business environments are being reinvented constantly, and corporate managers will need new skills and new tools in order to adapt. Changing economic realities will require better resource management, which, in turn, will generate systemic wealth by ensuring the optimal fulfillment of work in progress, both within and outside of the corporate organization. The capabilities inherent in enterprise networking is the obvious critical dependency in all of these developments.

The new enterprise webs and "virtual ventures" emerging throughout the 1990s will require managerial mind-sets that are global in scope and reach and yet grounded in local, hands-on realities that have remained elusive to "big systems" approaches. Rene Dubos's now well-known dictum "Think globally, but act locally" thus has its own set of resonances for a new style of corporate management that will rely heavily on new computer networking capabilities. The question in the future will center on not only whether managers meet the minimum acceptable criterion of being computer literate, but also how network savvy they are. Such skills will be indispensable in providing tools needed for the development of new communications-based marketing strategies. These strategies, as discussed in earlier chapters, involve developing radical new approaches designed to find the customer at the right time with the right "solution" (not just the right product or service), thus creating a whole new market dynamic.

To fully appreciate the role of enterprise networking in these scenarios, it's important to understand the full set of ramifications associated with the frequently invoked term "information-age economy." If Robert Reich is correct in his speculations that the job of symbolic analyst will become the dominant generic function in this new economy, then we must assume that information itself, in all its permutations, combinations, and transformations, will become the new currency that will be bought, sold, traded, updated, revised, repackaged, and then rebought and resold in different markets and different cycles. The value-added characteristics brought to information-as-currency will be imparted by the symbolic analysts who work with it on a day-to-day basis in their respective professional orientations. This holds true whether those professionals are working in the domains of science, education, architecture, computer science, mathematics, physics, or environmental engineering.

A NEW CONSTELLATION OF CORPORATE RELATIONSHIPS

Knowledge is power, but increasingly, knowledge is also wealth. In the years ahead, there is likely to be much more transparency between the academic and business environment, whereby ideas become valued not only for their intrinsic worth but also for what they can deliver in the real world, as translated into products and services that can improve the quality of life. Perhaps the commonly used phrase "it's academic" as synonymous with "it's irrelevant" will consequently fade into its own special oblivion. There are simply too many challenges to be faced in this new age of business and sociopolitical change for knowledge to go unused and remain in the silent preserve of education for education's sake. Academia may eventually come to be populated by individuals who can both "teach and do," and eventually universities may spin off more of their own think-tanks and consulting firms as did Stanford University with its Stanford Research Institute (SRI)—places where knowledge is applied to real problems, a kind of halfway house between academia and business.

At the heart of this alchemical process of taking knowledge-in-a-vacuum and facilitating its transformation into real goods and services is computer networking. Telecommunications in its broadest sense has shown itself to be a radical force that, with respect to academia, can help tear down the walls of the ivory tower. With respect to business, telecommunications can instill a new appreciation for both pure and applied research, building the knowledge base, and engendering new cooperative approaches to problem-solving. Networks such as the Internet-based SemNet in South Carolina (discussed in Chapter 7) are already providing the model for these new virtual communities of experts in a wide array of disciplines, coming together to create new structures and *ad hoc* enterprises. In the course of watching computer networking play out its capabilities, forward-thinking corporations should come to realize that much of their work will increasingly center on solving the world's many problems in a tumultuous age of change, where change itself is simultaneously aggravated and mitigated by the intelligent application of technology.

As various academic disciplines are gradually applied to the world of commerce and the development of goods and services, the potential exists for a new dynamic to take shape, the beginnings of which are already evident. With the value of new ventures increasingly acknowledged as dependent on the knowledge base that supports them, tasks, projects, and the "work of nations" will begin to gravitate toward the best specialized or generalized expertise available to tackle it. This stands in sharp and immediate contrast to the compartmentalization that the wrong kind of competition can engender. These new cooperative patterns, based on information sharing, will guarantee that badly needed human resources, especially in the scientific and engineering domains, are not shunted off into unproductive academic cul-de-sacs, but rather are brought into the agora of a dynamic new global marketplace where such skills are sorely needed to develop new solutions to new and old problems, many of them critical. Moreover, given the right path of applications development, the process of generating such solutions, could also serve to create profitable business ventures, while at the same time solving the massive difficulties that now beset nation after nation as the major issues of energy, environment, biological sustainability, and economic equity and stability are confronted. These new trends clearly represent a profoundly significant shift in the way in which enterprises are developed and engaged, a shift that in the long term, will likely also fundamentally alter the familiar economic structures that currently exist, even as they are already being eroded by the often abrasive machinations of rapid social and political change.

At the heart of such changes is the ongoing development of worldwide communications infrastructures, serving to create new channels of formal and informal communications between

- *Businesses and other businesses*, as evidenced by the now well-established trend of strategic partnering.
- *Businesses and individuals*, such that new arrangements are formed that transcend the one-dimensional structure of the traditional employer-employee relationship.
- *Universities and businesses*, allowing knowledge to become, where appropri-

ate, more fully engaged in real-world problem solving than has previously been possible, and also allowing corporations to become more socially oriented.

- *Individuals and universities*, which will create a new wellspring of independent study, supported and fostered by the university, but at the same time providing a far less constrictive environment for researchers wishing to pursue creative and innovative projects.
- *Universities and other universities*, whereby knowledge bases might be more readily shared, and via the Internet or other independent networks, made available for corporate use, as well as providing for their application in new entrepreneurial ventures.
- *Government and business*, allowing the highly competent engineering and scientific skill base in, for example, the defense community, to be applied and transferred to the solving of domestic and international problems involving energy, the environment, and the basic enhancement and refinement of computer and communication tools and resources; and finally, government with any or all of the above, as we begin to redefine the political structures that can accommodate this kind of change.

THE STRATEGIC VALUE OF INFORMATION

One of the most important features in the preceding permutations of new business relationships is the fact that computer networking is the enabler of a broad new set of communications channels that support them. Of course, relationships between these entities have always existed on one level or another. There is, for example, nothing new about businesses routinely communicating with other businesses or universities collaborating with corporations on specific research-oriented projects; or for that matter relationships between individuals and these organizations. *The change involved here is not one of degree, but rather represents a fundamental transformation of traditional modes of operation, based upon an emerging worldview that recognizes the need for new levels of cooperation between these entities in ways that were previously not possible.*

Similarly, when viewing the organization from an internal standpoint, the same phenomenon can be observed. In other words, the quantum change that's being described here applies with equal validity to the relationships that have been defined in Porter's value chain, as discussed earlier. New forms of horizontal, peer-to-peer communication made easily accessible to working groups throughout the corporate structure can allow fundamental working arrangements between those organization units to be renegotiated and restructured.

Information sharing, when allowed to take place between organizations that have not traditionally shared data, creates synergy. Synergy, in turn, creates wealth, in the purest sense. Information as a shared resource is also infinitely replicable, which makes its value increase as it passes to various functional entities in the value chain. (Such a capacity for infinite replication of course, is not without its own set of problems!)

I believe the ongoing massive exchange of knowledge and ideas will ultimately become a dynamic engine for the growth of this new global economy, based in large measure on virtual ventures and virtual communities, to use technology writer Howard Rheingold's phrase. Some readers might recall a science fiction movie from the 1970s called "Colossus: The Forbin Project." The movie depicted a day when the two major powers, the former Soviet Union and the United States, in a kind of metaenterprise networking event, hooked up their supercomputers and allowed them to exchange information freely and without restriction. In the long run, however, people, not computers, will increasingly be called upon to share their skills, experience, and knowledge, thus creating new and powerful synergistic approaches in the workplace. Computer networking simply represents the vehicle that will allow this to take place, if the right kind of infrastructures are built to allow it. Thus, strategic advantage, in large measure, will come first of all from providing the right environment for a new breed of symbolic analysts to carry out their activities; and second, from using these computer networking tools themselves to create new markets and to transform and optimize approaches already in place to address existing markets.

Such information exchange has been taking place on the Internet for many years, dating back to the days when the ARPANET was the primary enabler. What is different now is the scope, scale, immediacy, and ease of use of such exchanges as more and more personal computers have been deployed in both the home and the workplace. Worldwide networks, many of them still unknown and others uncharted, have strung their filaments across the globe like a spider's web. Autonomous networks such as USENET, BITNET, and many others have created new forums for ideas to be exchanged in a global "town meeting" to which increasing numbers of individuals have access. The fact that this is happening on an unprecedented scale in the university community constitutes the basis for much of the quasireligious fervor with which the Internet is discussed in those quarters. John Barlow, one of the founders of the Electronic Frontier Foundation, along with Mitchell Kapor, describes the development of these networks as "hardwiring our collective consciousness." What still seems to be lacking however, is a correspondingly enthusiastic sense of mission in the corporate world concerning the new levels of strategic advantage that are available when collaboration and cooperation—fostered by new forms of electronic communication—replace bureaucratic competition and compartmentalization.

THE NEXT MAJOR PARADIGM SHIFT: APPLICATIONS

As computer and communications technologies continue to grow and evolve, a major change in the way many executives, managers, professional specialists, and other symbolic analysts think about them will need to occur. Such a change would require both system users and designers to begin viewing the computer as an infinitely malleable tool, which, properly configured, can augment and enhance a wide range of business activities. The alternative is to view these fascinating devices as sophisticated surrogates for typewriters, calculators, and other types of mundane office equipment, which they certainly are not. (To paraphrase one executive's lament overheard during a flight from Boston to Los Angeles: "I've got the 1960 equivalent of a mainframe sitting on my desk and what do I use it for? As a fancy replacement for a typewriter!")

Thus, the major limiting factors in the development of state-of-the-art, strategically competitive applications have more to do with the "mindware" of planners, designers, managers, and users rather than any limitations imposed by computer hardware or software. New viewpoints and fresh approaches concerning the use of this technology are thus essential to the development of strategic computing tools. Executives must learn to become less computerphobic; managers must learn to empower their employees with these resources and resist the temptation to curtail their use for financial or overly cautious security reasons; and users must become more aggressive in applying computers and communications technologies to the everyday exercise of their routine and nonroutine business tasks, rather than waiting for the IS department to come up with a "silver bullet" that will do it for them.

Thus, for managers, one of the major messages of this chapter, and a major theme of this book remains: *empower users, because ultimately, there is no one in a better position to integrate applications and technology*. Given this fact, new roles for both IS and network managers will need to evolve, oriented toward end-user needs and applications and not toward technology. New staff positions will need to be developed with the intent of building easily accessible support networks for users. Having such resources will become an increasingly necessary requirement as more and more mission-critical applications migrate away from mainframes and onto client-server architectures, where downtime will be just as costly and damaging to business operations.

As these trends continue, IS itself will experience downsizing in several key areas, and many of its functions will continue to move in a decentralized fashion down to the departmental level. The widespread proliferation of LANs, along with the attendant internetworking infrastructures, has already created new requirements for departmentally based on-site LAN administrators, working either full-time or part-time, depending on both the size of the organization and the nature of the data traffic moving through it.

In general, there are a number of things that corporate-level IS and telecom staff can do to become more oriented toward users and applications rather than technology and equipment. Getting "out of the user's way" is the first dimension. At a minimum, this means developing a sensitivity toward possible impacts that companywide IS programs have on departmental users. Business is increasingly moving toward customized, individualized solutions. Poorly thought out and summarily imposed master plans that fail to take into account programs already established and working at the departmental level can create significant problems for users. Suffice it to say that getting to know the idiosyncratic and individual usage patterns of all the departments under IS purview is an important precondition to launching companywide programs.

Second, IS/telecom staff must reorient themselves as problem solvers and facilitators for less-than-computer-literate users. Doing this requires moving beyond comfortable familiarity with a specific vendor's products and services, toward those that are actually in use throughout the company. The point of view that many of these products and services have not been legitimately procured and therefore don't deserve support from corporate level IS is, needless to say, unrealistic and counterproductive.

Third, to the maximum extent possible, IS/telecom managers should become "advance scouts" for new and cutting-edge software and communications capabilities that are easily implementable on a corporatewide basis. Managers should encourage their staff to attend seminars, conferences, and trade shows in order to keep current with the latest technology and to explore new options that departments may or may not have the resources to look at themselves.

Finally, individual users and managers should be empowered via extensive training programs designed and specifically tailored to the applications they will be used for. All too often, this kind of training is generic in nature and, as such, is less than suitable for giving users the information they really need. This training should include informal sessions during which users are encouraged to discuss their real-world needs in a nonthreatening fashion so that education can proceed in both directions.

The other shift that must take place—and signs of this are already evident—is in the realm of software design: the need for what might be called a new type of software designer. Such professionals would need to have a completely different orientation toward the development of products, applications, and, most importantly, user interfaces than those currently in vogue. They would not simply be computer programmers with specific arrays of technical skills, but would have cultivated a more holistic appreciation for the human use of technology: the traditional realm of human factors engineering. Questions on typical usage patterns for computer and communications systems would need to be answered far more thoroughly in advance of product design than has traditionally been the case. For example, how do professionals *really* use e-mail systems? When are they most likely to gather their messages? What options do they tend to use more than others? We might call such individuals information designers, and there is every reason to expect that such positions will emerge not only among forward-thinking vendors during the middle of this decade, but also among users, presumably as part of IS/telecom management teams. Indeed, it's not inconceivable that users themselves, with the capabilities available via groupware, could become a dynamic part of the software design process, with specifications and product requirements being adjusted in real-time across communications links.

In the ideal scenario, information designers (some may prefer the term information architects) would be able to take as their *a priori* starting point a given set of users' application requirements. They would then help software programmers work "in reverse," designing programs and applications that are inherently and structurally user-friendly instead of attempting to add interface enhancements as an option or add-on as is now often the case. This will require collaborative teaming on the part of technical writers, customer support representatives, and field technicians who have real-world working knowledge of the problems and difficulties encountered by users of computer and communications software tools. This collaborative teaming will likely take place electronically, using the communications tools described in earlier chapters.

These factors are cumulative and interactive. Between the empowerment and education of users themselves, and emerging information design enhancements taking place at the user interface, the strategic use of information resources will phase-shift into a kind of exponential expansion and sharing of the knowledge base. This will, as Buckminster Fuller once predicted, create a new kind of wealth that will synergistically resonate, multiply, and perpetuate itself with the only prerequisite for success that it be shared and not partitioned, or allowed to become the fodder for new info-bureaucracies that could impede its fulfillment.

Many corporations have already learned the lessons of collaborative teaming and information sharing and, as a result, have become highly successful in their chosen fields of endeavor. Cisco Systems, a major router vendor for example, is known for its successful adoption of widely collaborative approaches in the marketplace; the company is extremely liberal and open with respect to the sharing of software code and applications with strategic partners and other market players. There are many other examples of companies that have learned these lessons, while the major computer-makers that have attempted to lock-in various segments of the marketplace via manipulation of their customer base and keeping a tight lid on their software designs have often languished. Thus, the idea of collaboration and cooperation is becoming part and parcel of the most successful of the newly emerging strategic approaches.

CHALLENGES FOR IS/TELECOM MANAGERS

If we can expect that users will enter a kind of information renaissance with respect to the use of these powerful new tools for the workplace, the technologies underlying them will also present a series of major challenges for IS/telecom managers looking to become flag-bearers for the new approaches. The transition from the big-systems approaches outlined earlier is one of these challenges, as is becoming more applications- and user-oriented. Another hurdle is the prospect of working around what might be called the reliability trap.

As computer and communications applications become increasingly mission-critical, there will be additional demands placed on IS/telecom managers to become more risk-aversive and conservative in their approaches to networking. Critical communications links that are not up and running will cost companies dearly in direct, on-stream revenues, often amounting to millions of dollars for every minute of communications link downtime. This being the case, senior management will increasingly look to these departments to become, in effect, the guarantors of reliable system performance—a major shift away from having a primary focus on strategic use.

The importance of these systems in the daily conduct of corporate business will only increase. It's not hard, then, to extrapolate that IS/telecom managers will be correspondingly tasked first and foremost with addressing such issues as disaster recovery, system backup and recovery, alternate path routing, communications route diversity, and so on. Obviously, in many instances, this will tend to narrow their focus as well as nudge it away from critical considerations involving user applications and technical support for those applications. With respect to telecom managers, such a trend might move their role backward into the all-too-familiar professional box of being the individual who makes sure that the phone lines are operating. On the other hand, various countervailing factors must be weighed into this scenario, such as the anticipated development of state-of-the-art network management tools that are automated, programmable in nature, and can, through the use of sophisticated AI and other techniques, off-load the baby-sitting and housekeeping chores of information resources managers to highly reliable self-governing network management systems. Until such time as "computers watching computers" becomes a more likely possibility, however, the principal management challenge will continue to be that of keeping one eye on user applications—where the real action will be—and doing whatever is necessary to ensure that system downtime is kept to an absolute minimum.

During the course of the next decade, companies will win or lose their market standing, based on their technological ability to develop new and more agile approaches to their traditional book of business. Fortunes will be made and lost as a result of preemptive first strikes carried out by visionary entrepreneurs and managers who can appreciate the simple logistics of information management:

getting the right information to the right person at the right time. As this book is being written, in the early 1990s, many companies have already come around to acknowledging the critical importance of the strategic use of networking; unfortunately, enthusiasm and forward-thinking management philosophies on the part of some IS/telecom staffers will not be enough. There is a decided knowledge gap that continues to exist in many corporations, and the keys to constructive change will continue to hinge upon such factors as downsizing and decentralization.

Downsizing will help companies break free of big systems thinking with respect to computer systems, and the traditional IS emphasis on expenditures, cost savings, staff budgets, and other typical corporate level considerations that were necessary in the older paradigm but are often characteristic of the bureaucratization of the IS function. We have seen how this bureaucratic approach toward the internal customer—along with an oversold and overoptimistic approach toward return on investment for large-scale systems—eroded confidence in IS approaches undertaken during the 1980s. PCs and PC LANs, along with groupware and powerful new options for desktop-controlled communications capability, will become the keys to making the use of network-based computing truly a force for change within corporate organizations. IS/telecom managers, therefore, must find a way to adapt their corporate roles accordingly.

Part of the new roles IS/telecom staffers will play in downsized and decentralized environments will revolve around facilitating and directing the process of change that is driving departmental computing toward greater independence. IS/telecom managers at the corporate level should begin to redirect their activities toward facilitating the transparent peer-to-peer connections of individual island infrastructures, incorporating wide-area connectivity into the picture, and constantly working to bring ease-of-use to the forefront. New internetworking technologies, based on "throwaway" black boxes, such as low-cost bridges and routers, will need to be smoothly integrated into the freewheeling environments characterized by downsized, easily deployable PC-based systems. Sound technical engineering can tie together both local- and wide-area communications in such a way as to provide easy access and low departmental charge-back costs, thus encouraging further use and exploration.

The word exploration is indeed a key here. "Power" users must be given the luxury of some trial and error with these new systems and the creative freedom to discover how they can best be applied to the constellation of their individual professional work responsibilities. Computers are not porcelain figurines. They won't break if pushed to the limit. They exist to be tested, experimented with, reconfigured, and put through the paces. This is a fundamental mind-set issue for managers and professionals who use computers, as well as for those who don't. A great divide clearly exists between those who see personal computers as inanimate objects that robotically carry out deterministic programming operations and those who view them as tools that can allow individuals to do whatever they ordinarily do better, faster, cheaper, and smarter.

IS/telecom managers, at both corporate and departmental levels, are in the best position to foster and facilitate such thinking. But if the right kind of individuals aren't in place to act as change agents for these goals, senior management must make the commitment to make the necessary changes. Furthermore, those managers must be given the leeway to develop the kind of training and education programs that will empower users and not reinforce the natural reluctance that most people feel when confronted by what appears to be an arcane type of know-how. IS and telecom managers must also try to communicate this new thinking to managers throughout the organization, while conveying the message that users should be allowed to experiment, and then share the results of those experiments with others via groupware and other team-oriented systems.

Other promising new areas of exploration relate to the use and design of the information flow itself. Barely an art or science, look for the design of presentation-layer information systems to become a major new trend in the 1990s. Developing these systems would ideally involve a kind of rigorous analysis of major parameters associated with task-based teaming. Based on that analysis, models of the types of information required for team-based project fulfillment would then be developed. I know of very few companies that perform this kind of analysis at present, but again, the major constraint that prevents this from being engaged more widely is the fact that most IS staffers are simply not trained to carry out such exercises. This can and must change. At a time when professionals and symbolic analysts move every day a little closer to a kind of permanent "condition red" of information overload, those who can design and manage systems that control and self-select the flow of information that's really needed will become highly valued members of the corporate staff and the new gurus of the information age.

Finally, new technologies should be introduced to those departments within an organization that may find a valid and valuable use for them. This is an important role that corporate-level IS/telecom managers can play, because they have a much wider familiarity with the range of vendors and available products in the market-place. An easy way to accomplish this is to arrange for vendors to come into the company and demonstrate their products for an assembled group of departmental managers, who are then free to individually contract with the vendor for a specific service. Taking this approach also ensures that incompatible systems will not be deployed throughout the company, and also gives IS/telecom greater visibility with those departmental managers.

As I have consistently argued throughout this book, the real work of developing new strategic networking approaches will take place at the desktop. Furthermore, it will be carried out by dedicated individuals who have learned new tricks in their old jobs using these powerful new technologies. The roles of IS/

telecom managers will increasingly emphasize facilitation skills, educational skills, and the ability to take the applications that have been creatively developed at the departmental level and, in so far as is possible, generalize them throughout the corporation. The continuing emergence of new technologies will accelerate and support these trends. Companies that adhere to these principles and emphasize the creative use of these powerful new tools will thrive in the new global enterprise environments of the 1990s and beyond.

It's critical to keep in mind that for corporations as well as for many other organizations affected by computer networking, much of this is uncharted terrain. Trial and error, a willingness to experiment, and an open-minded approach to what might seem to be radical or far-fetched ideas will serve IS/telecom managers well in developing the ability to transform computer and communications tools into genuine strategic resources.

Recommended Reading

Bradley, Stephen P., and Jerry A. Hausman (eds.), *Future Competition in Telecommunications*, Harvard Business School Press, Boston, MA, 1989.

Brand, Stewart, *The Media Lab*, Viking-Penguin, New York, 1987.

Crandall, Robert W., and Kenneth Flamm (eds.), *Changing the Rules: Technological Change, International Competition, and Regulation in Communications*, The Brookings Institution, Washington, D.C., 1989.

Donlop, Charles and Rob Kling (eds.), *Computerization and Controversy*, Academic Press, San Diego, CA, 1991.

Elbert, Bruce R., *Private Telecommunication Networks*, Artech House, Norwood, MA, 1989.

Huber, Peter, *The Geodesic Network: 1987 Report on Competition in the Telephone Industry*, U.S. Department of Justice, Washington, D.C., January 1987.

Keen, Peter G. W., *Competing in Time*, Ballinger, Cambridge, MA, 1988.

Opper, Susanna, and Henry Fersko-Weiss, *Technology for Teams*, Van Nostrand Reinhold, New York, 1992.

Reich, Robert B., *The Work of Nations*, Knopf, New York, 1991.

Runge, David A., *Winning with Telecommunications*, ICIT Press, Washington, D.C., 1988.

Savage, Charles M., *Fifth-Generation Management*, Digital Press, Bedford, MA, 1990.

Sproull, Lee, and Sara Kiesler, *Connections: New Ways Of Working in the Networked Organization*, MIT Press, Cambridge, MA, 1991.

Stone, Alan, *Wrong Number: The Breakup of AT&T*, Basic Books, New York, 1989.

Toffler, Alvin, *Powershift: Knowledge, Wealth, and Violence at the Edge of the 21st Century*, Bantam Books, New York, 1990.

U.S. Congress, Office Of Technology Assessment, *Critical Connections*, U.S. Government Printing Office, Washington, D.C., 1990.

About the Author

Thomas S. Valovic, editor of *Telecommunications* magazine, has been a member of its editorial staff for over five years. He was formerly Manager of Corporate Information and a speech-writer with Ameritech. He also held marketing-related positions with Motorola Codex and Racal-Milgo. His career began in 1973 as a science writer working on satellite communications projects at NASA's Goddard Space Flight Center. As a member of the client staff at the international management consulting firm Booz Allen and Hamilton, he worked in the areas of energy, environmental policy, and telecommunications. Mr. Valovic also served as a reviewer in conjunction with the U.S. Congress Office of Technology Assessment study on telecommunications, "Critical Connections." He has written hundreds of articles on the subject of communications technology and its business applications for *Telecommunications* and a variety of other industry publications including *PC Week*, *Computerworld*, *Information Week*, and *Communications Week*. Mr. Valovic received B.A. and M.A. degrees from Boston University in 1969 and 1970, respectively. He is on the adjunct faculty at Northeastern University, where he teaches courses in scientific and technical communications.

Index

The Artech House Telecommunications Library

Vinton G. Cerf, Series Editor